Safety Training That Delivers

How to Design and Present Better Technical Training

American Society of Safety Engineers ❖❖ Des Plaines, Illinois USA

Managing editor: Michael F. Burditt, ASSE
Editing, text design, and layout: Nancy E. Kaminski
Cover design: Publication Design, Inc.
Cover photos: Bill Cronin

Library of Congress Cataloging-in-Publication Data

Cantonwine, Sheila Cullen.
Safety training that delivers : how to design and present better technical training / Sheila Cullen Cantonwine.
p. cm.
Includes bibliographical references (p.).
ISBN 1-885581-29-7 (pbk. : alk. paper)
1. Safety education, Industrial. I. Title.
T55.2.C36 1999
658.3'82'0715—dc21 99-41073

CIP

Manufactured in the United States of America
First Edition
10 9 8 7 6 5 4 3 2 1

Dedication

This book is dedicated to my mother, Susan Cullen, who died from liver cancer. Ironically, we believe her illness was caused from a chemical exposure. I hope this book encourages more effective safety training in an effort to decrease future occupational illnesses and injuries. My mother spent her career dedicated to nursing, and I hope that this book will help many people avoid injury, disease, and premature death.

Table of Contents

Preface

In this book you will find many different training examples. Most examples are from actual experiences, although some are fictional and are used for illustrative purposes.

My primary experiences have been in the safety training and computer training fields; consequently, this is what I have written about. However, it is important to realize that most of the principles in this workbook apply to all types of employee training.

Acknowledgements

I would like to thank Ralph Allen and all my colleagues at the Office of Environmental Health and Safety at the University of Virginia, for introducing me to the safety field and for teaching me the "nuts and bolts" of chemical safety training. Thanks are also in order for Bruce MacDonald and the staff at North Carolina State University Environmental Health and Safety, for helping me to expand my safety training experience and for supporting me as I gained experience. I send a special thanks to Nelson Couch for encouraging me to start my own consulting business, and for helping me to write this material for a class I taught for the North Carolina Occupational Health and Safety Education and Research Center.

Thanks to Aruna Viswadoss, my friend and colleague, for reviewing my manuscript and offering helpful editorial comments.

I would like to thank my family and friends for their support and guidance. I especially want to thank my father, Anthony Cullen, who read my manuscript cover to cover and who offered many helpful suggestions. His love and support have helped me to accomplish everything I have always wanted in life. Thanks to Moira and Dan, my sister and brother, for a lifetime of believing in me.

Finally, and most importantly, I would like to thank my husband Paul for his love, support, and understanding. He was my sounding board and helped me formulate ideas for the creation of this book.

Chapter 1 Introduction

Have you ever attended a training class only to find that it soon puts you to sleep? It's probably not that the instructor didn't know the material or wasn't qualified. More than likely, the information wasn't presented in an interesting, effective, and stimulating way.

As safety and health professionals, we learn the technical aspects of our profession, but rarely are given guidance on how to effectively train others. In this workbook, I will provide you with insights and ideas that will enhance your training skills and help you to create more stimulating training.

I quickly learned while researching this topic that there is very little safety training information that is interactive. I know that I learn best by practicing what I am learning, so I have incorporated many exercises to reinforce learned concepts. This workbook can be used on your own or to facilitate a workshop with other trainers.

Over the years, I have trained more than fifteen thousand people in many different training situations. I learned a great deal from these experiences, and I hope that you will find new ideas as you read this book and work with the exercises.

Workbook Objectives

With this workbook, you will learn techniques for creating and delivering effective training courses.

This workbook will teach you to:

- Describe what constitutes an effective and ineffective training class
- Understand the adult learning process
- Conduct a training needs assessment
- Establish goals and objectives for training
- Develop curricula pertinent to the needs of the participants
- Explain the advantages and disadvantages of using different types of media
- Develop evaluations to use in training sessions

- Write appropriate questions for testing
- Know the most effective way to set up a training environment
- Comfortably answer questions
- Address various interventions
- Remember the characteristics of effective presenters

Characteristics of Effective and Ineffective Training

Before we start the workbook, take a few minutes to write down your thoughts. In your own experiences as a trainer and trainee, what makes training effective or ineffective? When compiling your list, consider both the design and delivery of training. See Appendix A for some possible answers.

Effective Training	Ineffective Training

Effective Training	Ineffective Training

As you can see, many factors determine effective training. What you have listed and what is in the back of the workbook are just a few.

The purpose of this workbook is to help you think of ways that you can make your training more effective. By making your training more effective, I guarantee you and your class will enjoy the training more. The participants will also increase their potential for retaining the information you present.

These are some questions that you might want to consider while you design and deliver training:

- What is the main point that I am trying to convey in the training?
- Am I presenting the information in an interesting way that stimulates the audience?
- Have I chosen an appropriate teaching method and proper audiovisuals?
- What audiovisuals should I use?
- Are my audiovisuals clear, and do they enhance or hinder my safety training?
- Have I chosen the best training environment possible?
- Is my training organized so it is easy to follow?

Chapter 2
Adult Learning

Why Adults Learn Differently

One of the most important things to remember with adult learning is that we learn things differently as children and adults. Most of the learning that we experience during our lives occurs while we are children. When it comes time for us to teach, we think about the ways that we were taught as children. This can be a bad way to approach the learning process, because adults learn things very differently than children.

To illustrate these points, let me give you some examples. In addition to teaching safety training classes, I also teach computer classes. Before I started a recent class, I asked everyone to explain why they were at the training and what they hoped to learn or gain. One woman explained that she was there to learn some training techniques so that she could teach the same class next week at a local community center. The audience was going to be teenagers.

At the end of the class she approached me and asked for some pointers on how to teach the class. One of the assumptions she had was that the pace should be slower since her audience would be younger. She thought it would be more difficult for the teenagers to learn the software because they have less experience with computers than adults do. I explained that it was the exact opposite. I recommended that she pick up the pace for a younger audience because their minds are constantly in the learning mode. Most of the teenagers were probably also very familiar with computers, whereas adults often fear and avoid them.

What's the bottom line? You have to remember that there are certain factors that cause adults to learn differently than children. Following are some facts that you need to consider when teaching adults.

Adults have many years of experiences that affect their learning ability.

You may be teaching a chemical safety class for a laboratory. You may have someone in that audience with a Ph.D. and you may have someone who has not completed high school. It is safe to assume that they will each feel differently about learning new materials. The person with the Ph.D. would feel comfortable, whereas the individual with limited education might feel anxious in a training environment. Always think about the different learning experiences that people may bring to the training.

People may also have different work experiences that may change their attitude about learning as well. I once went to a department where there had recently been a bad accident. One of the employees had accidentally chopped off the end of his finger with a paper cutter. It was a very serious accident and most people were very worried about their coworker and the hazards posed by the paper cutter. Because of the accident, this department considered office safety a serious concern. When I went to talk about office safety, there was a definite respect when I talked about the dangers of using a paper cutter.

Others who had not seen the ramifications of using an unsafe paper cutter were not as eager to listen to the various hazards you could face. Since this was not the first time that this had happened in the organization, we had incorporated a slide of a paper cutter and some information about safe use in our new employee orientation. The majority of the time, people smiled and almost snickered when I explained how dangerous a paper cutter could be. It was very different than the training for the department that had experienced one of their colleagues losing the tip of his finger.

Adults have a "reality filter."

A "reality filter" allows employees to filter out suggestions that are realistic while ignoring those that are not. Basically, they are able to determine if what you are advocating in the safety training is realistic for them in their job. For example, you may be conducting a class on Lock Out Tag Out. In your safety training, you explain the importance of locking and tagging electrical equipment before maintenance is performed on the equipment. The reality filter may tell them that they won't have time to mess around with all that "safety stuff" when they have to meet production deadlines.

Adults almost always consider what you say in terms of what they can and cannot realistically do. People know their jobs better than any instructor does, and they are constantly determining how your policies will realistically work when they also have to consider time, production, and monetary constraints.

Time constraints are perhaps the most common reason that people use their reality filter to argue that they won't be able to implement safety procedures. I remember a situation where employees were using back belts incorrectly in their jobs because they felt it took too much time to use them correctly. In an effort to save time, they were wearing their back belts tightly fastened all day long instead of wearing it loosely, then tightening it only when they were lifting. When the supervisors explained the correct way to wear the belt, many employees complained, saying they didn't have the time to make the belt tight and then loose before and after every lift they performed. Time constraints can be very real.

Aggressive production schedules can also cause employees to use their reality filter. When management expects a certain volume of material to be processed, employees do what they can to make that happen, often at the expense of safety. Lock Out Tag Out is a good example of this. Let's say that the employees are running a long assembly line. One section of the line breaks down and needs some maintenance. The safe action would be to lock and tag out the energy sources for the assembly line. Doing this would stop production. The fear and uncertainty caused by missed production schedules may tempt employees to work around Lock Out Tag Out procedures.

Finally, monetary constraints can cause employees to use their reality filters when it comes to implementing safety procedures. This is most often a problem with personal protective equipment (PPE). Although the hazards may suggest or require the use of PPE, monetary constraints may prevent the employee from having the proper equipment, or the equipment they have may be old and ineffective.

A good example of PPE and monetary constraints is the use of dosimeters in radiation protection. I heard of a case where all employees that were in the vicinity of radiation were issued a dosimeter no matter how low their dose or risk to radiation exposure was. It was determined after some time that the program was too costly. After evaluating dosimeter readings for the past years, it was found that many jobs repeatedly had no recorded exposures. Therefore, in an effort to save money, those employees had their dosimeters taken away. In the long run it was the correct decision to make because the radiation protection department may have been overly cautious. Money drove the change to decrease the number of employees participating in the dosimeter program.

As you can see, reality filters can be a very powerful influence concerning safety training. What the instructor says in class and what the employees can realistically control are sometimes very different. The trick is to recognize that reality filters exist and consider them in your training.

A good approach with safety training is to constantly remind your audience that the whole purpose of the training is to help them do their job more safely, which ultimately protects their health and well-being. Remember the saying "what's in it for me?". That will help you to relate to the reality filter that adults use. Try to point out that safety procedures benefit them. They may inconvenience them in the short run, but they will benefit them in the long run. Remind people that a disabled or dead employee has no productivity or future career.

- **Adult learners have set habits and ideas.**

  If class participants feel that safety training is a waste of their time, and they have felt that way all their lives, it may be difficult for you as the trainer to overcome this idea. You may not be able to change their opinions but you do need to recognize them.

  Realize that habits can also be hard to break. For example, it used to be acceptable laboratory practice to pipette chemicals by mouth. We have learned that serious accidents can occur when using your mouth to pipette, so people have had to learn to pipette with a bulb instead. After having used your mouth for years, you can see how it would be hard to change. It is a lot easier for new laboratory workers to pipette with bulbs since they have no previous habits to overcome.

  A more practical example would be the use of seat belts in cars. Many of us learned to drive when seat belts were not required and in some cases not even available in cars. Research and studies have indicated that wearing a seatbelt can reduce your injuries in an accident and may even save your life. Because of that research, we have all had to change our habits to include buckling up before we start the engine. Car manufacturers even help by creating cars that have automatic seat belts. They recognize old habits are hard to break.

- **Adults learn more effectively when they know that the information is relevant to their jobs.**

  I remember giving a chemical safety talk to office staff because solvents were found in correction fluid and markers. My goal on those occasions was to teach some basic chemical safety. I would have lost the class' attention in a minute if I had started teaching them about efficient chemical inventories and proper chemical labeling, because that did not pertain to their jobs. Always try to make training relevant to each specific group you are training.

- **Adults experience a decrease in vision and hearing ability.**

  This is sad but true. The American workforce is getting older, and we need to take predictable effects of age into consideration when we design our training. Make sure that your visuals are created with large fonts and that your voice projects. If you can not connect to people with visuals and your voice, then you have little chance they will listen to or understand your training.

  One of the biggest problems I experienced in teaching computer classes is helping people who have bifocals. I noticed they generally had a hard time keeping up with the rest of the class. No matter what I did or how slow I made the class, it seemed they were always behind. After standing behind one student, I finally realized the problem was that she had bifocals. She was not able to clearly look from her screen to her book to her keyboard and then to the projection screen. She was not able to see unless she continually put on and removed her glasses, and that took a great deal of time. Having dim lighting for the projection screen certainly did not help. After making this realization, I tried to be more sympathetic to people who do not have perfect vision by giving them a little more time.

- **Finally, the ability to learn new concepts and skills requires application to the job.**

Most people learn better when they are actually doing something rather than just hearing information explained to them. For example, when you are giving training for accident investigations, wouldn't it be more beneficial to have people actually perform a mock accident investigation? They could look back on that first experience with confidence when it comes time for them to do an actual accident investigation. When they leave the class they have already experienced the dreaded "first time."

As you can see, there are many differences between teaching children and adults. Adults have different experiences, set ideas and habits, a reality filter, and they experience a decrease in hearing and vision. Adults also learn best when the information is relevant to their jobs, and the ability to learn new concepts requires practice on the job. It is important to recognize these differences when you are creating and conducting your training classes.

"Adults have a reality filter."

"Adults have previous Experiences."

"Adults learn more effectively when the information is relevant to their jobs."

"Adults experience a decrease in vision and hearing."

"Adult learners have set habits and ideas."

"Habits can be hard to break."

"The ability to learn new concepts and skills requires application on the job."

Eight Principles of Adult Learning

Here is a more formal way of looking at some key principles of adult learning:

Principle	Explanation	Suggestions for Application
Readiness	Adults learn only when they are ready to learn and can see a reason for learning.	Motivate the participants early. Explain what they will learn, how they will use the knowledge/skill, and why it is important to them.
Application	Adults learn best by doing, that is, when they are active, rather than passive participants in the learning process.	Strive for the 30/70 goal in presenting training: 30% lecture and 70% activity/involvement.
Context	Adults organize information (content) internally based on how the information is to be used (context).	Stress how knowledge/skills will be used. If at all possible, deliver the content in its job context.
Exercise	Things most often repeated are best remembered.	Plan for repetition of critical knowledge/skills.
Effect	Learning is strengthened when accompanied by a pleasant or satisfying feeling; it is weakened when associated with an unpleasant feeling.	Give positive feedback. Praise/reinforce success. Minimize criticism.
Primacy	What is often stated first creates a strong, unshakable impression.	Present correct information the first time.
Intensity	A vivid, dramatic, and exciting learning experience teaches more than a routine and dull experience.	Use training media that appeal to as many of the trainees' senses as possible.

Principle	Explanation	Suggestions for Application
Recency	All things being equal, the things last learned are first remembered.	Repeat all critical knowledge/skills that may have been forgotten because they were presented first in the training. Summarize often: important content should be reviewed at the close of each period/day.

Source: Rebecca Bullard, Mary Jean Brewer, Nancy Gaubas, Angie Gibson, Kathy Hyland, and Eileen Sample, *The Occasional Trainer's Handbook*; Englewood Cliffs, NJ; Educational Technology Publications; 1994 (4–12)

Discussion of the Eight Principles of Adult Learning

You need to think about the eight principles when you are designing and delivering your training. Let's consider each of the eight principles in more detail.

Readiness

The key word here is motivation. Adults learn most effectively when they see a reason to learn. Therefore, you need to quickly tell them what they will learn and how it will benefit them.

When I teach computer software classes, I always ask people why they are there. Since the classes are voluntary, most people come up with some interesting projects that they are working on. For example, they may say, "I want to learn how to create web pages so I can spend time maintaining my department's web page." Or they may say, "I want to learn about databases so I can keep track of all the employees, where they live, where they work, and when they get paid." The point is that they are not only ready to learn the information but that they have a need to learn.

In some cases the participants may not have their own motivation, so you must supply it. The best way to do this is to explain within the first minutes the reason why they are at the training. Standing up and saying "Okay, we're all here because OSHA makes us" just doesn't cut it. Many people think that money is what motivates people in their jobs, when in actuality life enrichment is what really motivates them. If you tell them they will have a more fulfilling life by avoiding workplace accidents, that is usually enough motivation.

Application

People learn by doing something. Wouldn't you rather be active? I know I hate going to training sessions where I sit in my chair for hours on end. Creating activities not only holds people's attention, it also helps them to learn.

When I originally developed the class that this book is based on, I tried to create an activity for every concept that I taught. My goal was to spend roughly a half-hour on each topic with ten minutes on lecture, followed by twenty minutes of activity. The people in the class enjoyed the activities and it made it easier for them to remember the information.

There are two hidden benefits for you, the instructor, when you follow this approach. First, you give yourself a break from talking. I have never had to teach for eight hours straight without any activities, but I can imagine it would not be much fun. I doubt you would have any voice left after 480 minutes of constant talking.

The second benefit is that you allow your audience to teach you new information. For example, I start my class on training techniques by covering what constitutes effective and ineffective training. Even though I have a rough list of my ideas, I usually ask the class to break into groups, discuss their thoughts, and then report back to the class. I am amazed at how many different responses they think of, some of which I have never realized.

Context

For this principle, you want to remember that people organize what they are learning based on how they will actually use it. More simply stated, they are putting the information in context.

As the instructor, your best plan is to emphasize how the information will be used in their job. For example, if you were teaching a class on how to use a fire extinguisher, it would be beneficial to cover information on how fires might start. In most cases you might only be expected to discuss the different types of extinguishers and what types of fires they work with. Although you would technically be covering the required information, some people might think to themselves, "I can't imagine ever having to put out a fire in MY job" and they might not pay as close attention as they should. But, if you explain different scenarios that would warrant using an extinguisher for different types of fires, then all of a sudden they are able to put the training in context.

Using examples and telling stories are excellent ways to help people put the information in context. Learning from other people's mistakes and accidents can be very valuable in training.

Exercise

Repetition is the key word for this principle. If you want people to remember something, repeat it as often as possible.

For example, when I conducted new employee safety training, it was important that all new employees knew the emergency number. Dialing this emergency number would provide medical, fire, and police assistance so it was vital they remembered the number, especially in stressful emergency situations. Although the number was relatively easy to remember, I made sure that I repeated it at least four times before the end of the training session.

Effect

Be positive! There are many different ways you can exude a positive attitude when conducting the training and while interacting with the participants. For example, always give positive feedback, even if their response is not correct. Being positive encourages people to get involved, while negative feedback will intimidate the group.

Also try to minimize criticism, not only of the people in the class, but also of other people or organizations that you discuss (e.g. management or OSHA). I've heard instructors start classes by saying things like, "We all know that OSHA is out of touch and is out to get everyone in our industry, but we still need to go through this stupid training." We can all agree this is a terrible way to start a training class. Criticism can threaten people because they never know when the criticism is going to be turned on them. If you are tempted to criticize, think of the old saying; "If you can't say something nice, don't say anything at all."

Primacy

People are usually most alert at the beginning of training, so make sure you present the most important information first. Think about watching the evening news. Although they may have many stories to cover, they always open with the most important ones. They realize they will lose their audience if they do not get their attention immediately.

Intensity

Be exciting and energetic. No one wants to go to a dull training class. The experience will be much more enjoyable and people will learn more when the training is memorable and exciting. More information on this can be found in Chapter 3, "Designing Effective Training," and Chapter 5, "Delivering Effective Training."

Recency

This is similar to the fourth principle, in that both principles involve repetition. Recency is a little different in that it is specific to the end of the training or the end of a training segment. Even though you may have stated the key information in the beginning of the training (Primacy), and you have repeated it throughout the training (Exercise), you will also want to review it again at the close of each training segment. If you are teaching a class that may extend more than one day, it is good to summarize at the end of each day. Even though people have "information overload," reminding them of the key points at the end of the training will help them to retain the information.

Exercise •

Adult Learning Exercise #1

Now that we have covered some different principles for adult learning, we are going to switch gears and talk about how people remember information most effectively. Knowing how people learn helps you design the training—knowing how they remember your information helps you deliver it.

Following are two exercises. Read both stories and then answer the questions that follow. In the first exercise, read the story to yourself only once and then answer the questions that follow. In the second exercise, read the story out loud as you write it and then answer the questions that follow. The purpose of these exercises is to see which story you remember better and why.

Read the following story to yourself and then answer the questions that follow *without* referring back to the story. It may be tempting to look back at the story, but it's important that you read the material only once before answering the questions. The object of this exercise is to see how much information you retain after reading the story to yourself only once.

The Story...

Bob came to work Tuesday morning ready to start the day. He has been working for the same company for the last seven years. He was a little distracted this morning because his eight-month old daughter was fussy, so he forgot his brown steel-toed shoes. He was in a hurry after that so he forgot to tie back his blond hair with his blue rubber band. He also had the misfortune of having his 1995 red pick up truck break down on the way to work. He was clearly distracted when he finally arrived at work. As he prepared to operate his table saw, he realized that he was still wearing his white gold wedding band and his emerald high school class ring. Even though he was still distracted from the morning's events, he had the good sense to make sure that he put on his spare pair of black steel-toed shoes from his locker. He also remembered to tie back his hair and remove his two rings before he started the saw at 9:45 a.m.

Now turn the page and see if you can answer the following questions without referring to the story.

Exercise •

Questions for Adult Learning Exercise #1

_________________What is the name of the main character?

_________________What type of vehicle does he drive?

_________________What kind of saw does he operate?

_________________On what day of the week did the events in this story happen?

_________________How long has he worked for this company?

_________________What time did he turn the saw on?

_________________What color is his spare pair of steel-toed shoes?

_________________What stone is in his high school ring?

_________________How old is his daughter?

_________________What color is his hair?

Now, go to Appendix B to check your answers with the correct answers.

How many did you get right and wrong?

________Right answers

________Wrong answers

________Percentage right

Generally speaking, people remember 10% of what they read. If you scored high for this test, come back tomorrow and try to take the test again without reading the story. Then come back in a week and in a month. Chances are that you will remember less and less as time goes on. This is what happens when training is in a written format only. People may remember it the first day and even do well the first week, but after that they may retain very little of the information that they read.

Let's see how the percentage of correct answers can increase when you say and write the story, thus increasing your interaction.

Exercise •

Adult Learning Exercise #2

Write the following story in the space provided and say the words out loud as you write them. Then answer the questions without referring to the story. The object of this exercise is to see how much information you retain when you hear and write the story.

The Story...

Marianne got a new office with a new computer, monitor, and chair. The first thing she did was set up her new computer. Her CPU (Central Processing Unit) was a tower model, so she decided to put it on her new gray carpeted floor. Next she set up her seventeen-inch monitor. She thought about putting it in front of the window, but then she thought the bright light might bother her eyes. The sun shines a lot in Georgia. She had to bend her neck back to see the monitor so she decided to raise the seat of her chair by three inches. Luckily, her chair had a lever on the bottom left side so she could easily adjust the height of the seat. She could also adjust the height of her back support. Then she moved the keyboard and trackball. She wanted to make sure that her arms rested comfortably and that her wrists stayed in a neutral position. Finally, she moved her black footrest to complete the new office.

Now, rewrite the story, saying the words out loud.

Exercise •

Questions for Adult Learning Exercise #2

After you have written the story and said the words out loud, answer the following questions without referring to the story.

_________________What color is the footrest?

_________________Where did she first want to set up her computer station?

_________________How big is her new monitor?

_________________What state is her office located in?

_________________What did Marianne set up first?

_________________Where was the lever on her chair?

_________________Where did she put the CPU?

_________________How high did she raise her chair?

_________________What color was her new carpeting?

_________________Does she have a track ball or mouse?

Now, check your answers with the correct answers in Appendix C.

How many did you get right and wrong?

________Right answers

________Wrong answers

________Percentage right

Generally speaking, people remember 70% of what they say and write. Chances are you scored higher on the second exercise than you did on the first. You would probably see an even larger difference if you were to come back and repeat the tests after some time has passed.

Why was the second story easier to remember? Because you actually did something to help you remember—you wrote the story down. To reinforce it even more, you heard yourself speak the story. Involving more senses helps you to remember.

As you can see, you can remember concepts better when you use many senses. An easy way to remember this is by looking at Dale's Cone of Experience.

Dale's Cone of Experience

Verbal Symbols

Visual Symbols

Radio, Recordings, Still Pictures

Motion Pictures

Educational Television

Exhibits

Study Trips

Demonstrations

Dramatic Participation

Contrived Experiences

Direct Purposeful Experiences

Source: Edgar Dale, *Audiovisual Methods in Teaching*, third edition; Hinsdale, IL; The Dryden Press Inc; 1969 (107)

In interpreting this, you can say that people remember very few verbal or visual symbols but they have a greater chance of remembering a dramatic participation or direct purposeful experience. Another way of interpreting this is to say people usually remember:

- 10% of what they read
- 20% of what they hear
- 30% of what they see
- 50% of what they hear and see
- 70% of what they say and write
- 90% of what they say as they do a thing

What are some practical ways that you can use the ideas of Dale's Cone of Experience in your own safety training? Remember, the more active the person is in the learning experience, the better chance he or she will have of retaining the information. It is even more beneficial if you can make the learning fun. People forget that they are learning when they are having fun, and it makes training a more pleasant experience for you and the participants.

There are many different ways you can make training interactive. My favorite is the use of games. With games you can actually do something rather than just see visuals or hear the instructor's voice. Games can also be a great way to test a person's information retention.

There are different games that you can play, but my favorites are crossword puzzles and "safety jeopardy." For crossword puzzles, you need some words related to the topic and the associated definitions. I usually use software to create the puzzles. You can find crossword puzzle software applications at most office supply stores. I like to use crosswords at the end of training so they are used not only as a fun activity, but also to reinforce information retention. I ask the group to form teams so that people do not feel unnecessary individual pressure if they are not comfortable working with puzzles.

The teams try to complete the puzzles, and the first team done usually gets a prize. I almost always use candy as a prize. It sounds childish, but who wouldn't want some chocolate after a hard day's work? After the winners have been congratulated, find reasons why the other teams should also get candy. Look in Appendix D for some examples of crossword puzzles that you could use in your training. There are examples of chemical safety, laser safety, and hearing conservation crossword puzzles.

Safety jeopardy is also one of my favorite games, although it can be more difficult to create. For safety jeopardy, you need to have different questions with varying degrees of difficulty, so different point values can be assigned. For example, a 100-point question for laser safety may be "The most dangerous class of laser." The answer would be "What is Class 4?" A 500-point question may be "The three properties of lasers." The answer would be "What is monochromatic, directional, and coherent?"

Once you have assigned the questions and their point values, all you need to do is put the points on a flip chart or an overhead so that everyone can see the categories and the points. I usually have four or five categories along the column headings and then three different point values (100, 200, and 300) along the row headings. See Appendix E for an example of Ergonomic0s Jeopardy.

Once the chart is completed, all you need to do is host the game. Normally I have all the questions and answers written on a piece of paper in front of me to avoid searching for the correct answers. Then you create two or three teams, depending on the size of the audience, and start the game. One team picks a question, "Saws for 200 points," and then both teams compete for the answer. The points are assigned to the team that answers the question the fastest. Once a team answers a question correctly, they can then choose the next category.

If you wish, you can also have "daily doubles," where teams get to determine how many points they want to wager for each question. You can also have a final jeopardy, where the teams wager points for the final category. All in all, it is great fun, builds teamwork, and helps the participants remember the key training points.

These are just two of the games I like to use for safety training. There are many others. It does not take a lot of time to create these games, and the benefits are enormous. Word will spread that training is fun, and people will have a better attitude about going to mandatory training. You also increase the effectiveness of your training, because people will remember the information.

There are other ways that you can make your training interactive. See Chapter 3, "Selecting a Training Method," for more information on ways to involve the audience.

Memory and Thinking Styles

Next, you should consider the different ways that adults think and remember things. Generally speaking, people are either auditory, visual, kinesthetic, or a combination of these.

- Auditory learners:
    - Remember in terms of what they have heard
    - Learn best by listening or having someone explain something to them
    - Use language: "That doesn't ring a bell" or "It sounds good to me"
    - Speak more slowly, and their speech is often rich with inflections and intonation

- Visual learners:
    - Think, imagine, and remember in terms of pictures and images or things they have seen
    - Speak with visual references: "I see what you mean" or "I can picture what you're saying"
    - Frequently talk in a high pitch at a fast rate

- **Kinesthetic learners:**
  - Learn and remember through feelings and tactile sensations
  - Learn by doing
  - Use language like: "I'll be in touch" or "I need to get a feel for the material"
  - Have the slowest speech patterns

Remember your participants may favor one of the three styles or may use a combination of all three. It is important to appeal to all styles in your training. For example, they can hear your voice as you speak (auditory), they can see your slides, overheads, or videos (visual), and they can actually practice what they learned in group activities (games, discussions, or role playing).

The following test by Dr. Donald J. Lofland will enable you to determine your own memory and thinking style.

Testing Your Memory and Thinking Style

by Donald J. Lofland, Ph.D.[1]

Following are some questions to determine how you think and remember. For each question there are three answers:

3 = your best choice
2 = your second choice
1 = your last choice

I most effectively communicate what is happening inside of me

_____ a. through my tone of voice and choice of words

_____ b. through my eyes

_____ c. through my posture and the emotions I convey

When I don't quite understand or remember something

_____ a. it doesn't ring a bell

_____ b. it seems hazy or unclear

_____ c. I can't get a feel for it or get a handle on it

I most easily notice

_____ a. the quality of music from a stereo

_____ b. if colors or shapes clash

_____ c. if clothes feel uncomfortable

When I am fully involved, I am _____ what I am doing

_____ a. tuned in with

_____ b. focused on

_____ c. in touch with or connected with

I express myself best by

_____ a. speaking my ideas

_____ b. describing my picture or vision

_____ c. writing my thoughts or expressing my feelings

---

1. Source: Donald L. Lofland, *Powerlearning—Memory and Learning Techniques For Personal Power;* Stamford, CT; Longmeadow Press; 1992; (57–65)

In a discussion, a person will most likely get my attention by

______ a. their tone of choice and choice of words

______ b. their point of view

______ c. their emotional expressiveness

When I have leisure time I prefer to:

______ a. listen to music

______ b. sightsee

______ c. go dancing

An effective way for me to make decisions is to rely on

______ a. what sounds best to me

______ b. what looks clearest to me

______ c. gut-level feelings

Learning academic or technical materials is easiest for me when

______ a. someone explains the ideas to me

______ b. I can visualize the concepts and see the whole picture

______ c. I can learn by doing or get a feel for the ideas

At a party, I am most attracted to people who

______ a. are interesting, articulate speakers

______ b. radiate visual beauty

______ c. convey a warm, relaxed feeling

The most important thing to help me to remember directions is to

______ a. repeat them to myself as I hear them

______ b. visualize them

______ c. write them down or intuitively sense how to get there

Once I completely understand a new idea or concept

______ a. I have it loud and clear

______ b. I can envision it

______ c. I have a feel for it

In my environment I am most likely to notice

______ a. sounds and quietness

______ b. visual coordination and beauty

______ c. feelings of warmth and/or comfortable furniture

Memory and Thinking Style: Scoring the Test

Once you have finished taking the test, add up your scores.

Sum of numbers next to "a" responses ____

Sum of numbers next to "b" responses ____

Sum of numbers next to "c" responses ____

"a" responses correspond to audio
"b" responses correspond to visual
"c" responses correspond to kinesthetic

Based on my responses, my
predominant memory and thinking style is ___________

My secondary memory and thinking style is __________

My lesser memory and thinking style is______________

It would be ideal if you could have all your participants take this learning style test before you conduct your training, but that is not normally realistic. What you want to remember is that everyone has different learning and memory styles. You need to respect that difference and appeal to all three styles when you are creating and conducting your training.

It is also beneficial for you to recognize your personal learning style so that you can experience training sessions more effectively yourself.

Chapter 2 Summary

As you can see, there are many different aspects of adult learning. For example, adults have previous experiences, they have reality filters, they have set habits that may be hard to break, they need information to be relevant to their job, they experience a decrease in hearing and vision, and they need to apply new concepts and skills. Remembering these ideas will help in creating and delivering effective training.

When training adults, you need to also consider the Eight Principles of Adult Learning (Readiness, Application, Context, Exercise, Effect, Primacy, Intensity, and Recency.) These principles basically say that you need to motivate adults, get them involved, put training information in context, repeat key concepts, be positive, present the most important information first, be energetic, and summarize often.

The adult learning exercises found in this chapter reinforce the concepts taught by Dale's Cone of Experience. Increasing the number of senses people use will increase information retention. Try to involve people in as many ways as possible. Games like crossword puzzles and jeopardy are excellent ways to involve people in training.

It also helps to be aware of your memory and thinking style. Although people may favor one of the three styles (auditory, visual, or kinesthetic) you as an instructor need to appeal to all three styles.

In conclusion, there are many different things to remember when teaching adults. The entire chapter can be summarized with these key points:

- Adult learning is affected by previous experiences
- Knowledge and skills need to be relevant to people's jobs
- Interaction increases information retention
- Involving different senses (sight, hearing, touch) increases information retention
- Repeat critical information and summarize
- Have a positive attitude
- Incorporate all three memory and thinking styles

Chapter 3
Designing Effective Training

Up to this point, we have discussed what makes safety training effective and ineffective. We have also thought about some different aspects of adult learning. We have learned that training is more effective when it is interactive, and we have even determined our personal memory and thinking styles. Now it is time to design some training. Below are the steps that we are going to take to create effective training:

A. Completing a needs assessment

1. Types of needs assessments
2. Getting management to support a needs assessment
3. Training needs indicators
4. Needs assessment survey

B. Establishing goals and objectives

1. Types of objectives
2. Helpful verbs for writing goals and objectives
3. Creating goals and objectives

C. Gathering information

1. Where to gather information
2. Selecting a variety of material

D. Organizing your training

1. Using a needs assessment to create objectives and a training outline
2. Training outlines
3. Presentation patterns

E. Training methods

1. Selecting a training method
2. Advantages and disadvantages of using different training methods

F. Using audio-visual aids

1. Example of a good and bad visual
2. Using different types of media
3. Pros and cons of different types of media
4. Ranking training on seven dimensions

G. **Testing and developing questions**
   1. Helpful hints to remember
   2. Guidelines for written tests
   3. Examples of well-constructed test questions and poorly-constructed test questions

H. **Logistical arrangements**
   1. Facilities, room arrangements, and equipment set-up
   2. Seating arrangements

I. **Trainer and audience evaluations**
   1. Why do an evaluation?
   2. Creating effective evaluations
   3. Sample evaluations
   4. Don't forget your own evaluation

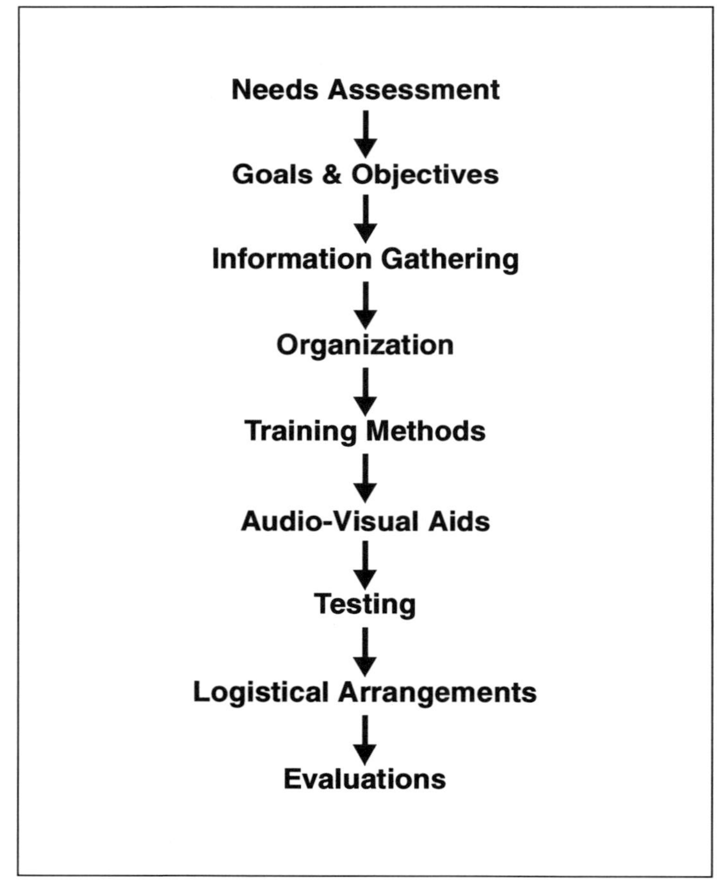

Needs Assessment

The first step in designing effective training is to complete a needs assessment. What is a needs assessment? For those of you not familiar with this term, it means asking questions to determine what causes a need for training.

Needs assessments can be very valuable tools in determining important information about the audience. You are deciding if they need the training. Furthermore, you are determining more specific information about what training they need. For example, do they need a brief overview, or do they need a detailed and complete explanation of information?

I perform needs assessments because it helps me to better prepare for the training. I hate that sick feeling you get when you begin training and quickly realize that you have prepared the wrong information for the audience. It's better to find out their needs before you start the training rather than try to "wing it" as you change your presentation on the fly.

Here is an example of winging it. I was asked to do a combined training session on chemical and radiation safety for employees who worked in buildings with laboratories. I had prepared a half-hour presentation on each topic with some activities at the end. What I did not realize was that they had changed their mind and decided to send ALL employees to the training, regardless of where they worked. They had originally told me that only people working in labs would attend the training. The reason for extended training was that people needed to be quickly transferred to various buildings when other employees were sick or on vacation. It was possible that all employees might have to work in labs occasionally, and therefore they should have the training.

The change made sense to me, but it caught me completely off guard the first time I conducted training. I started the training by asking people to introduce themselves and where they worked. After hearing from all twenty people I quickly realized that none of them had even seen a laboratory. It meant I had to change the way I went through the training and I had to help them with the activities in the end. Had I known that the audience would be mixed (lab and non-lab personnel), I would have included more detail for those employees not familiar with labs.

Other problems arose during those sessions. I started realizing after the third or fourth session that the audience seemed restless during the training, and some of them brought food. The training was at 9:30 a.m. so I thought it was odd that they would be eating what appeared to be their lunch. I later found out that they started their shift at 5:30 a.m. and their lunch break started at 9:30. Although I don't advise having training during a lunch break, it would have been beneficial if management had told me that it was their lunchtime.

Well-constructed needs assessments can provide a large volume of helpful information. For example, they can tell you:

Who needs the training?

Is it for all the people in a department or just a few? There may be some employees who have already had the training recently, so you may only have a small group of people who need it. It is also good to know exactly how many people to expect so you have enough materials for everyone. Other questions you may want to ask about your audience include:

- What is their knowledge and skill level?
- What are their attitudes?
- Do the people work in the same department or are they from different locations?

Who requested the training?

You might approach the training differently if employees rather than management requested it. I tend to ask more questions if management requests the training, because they can sometimes be unaware of the people's needs. I also make sure I spend some time talking to the employees before the training, regardless of who requested it.

What is the educational level of the people who will need the training?

It is important to recognize how much skill and knowledge your audience has. The more you know about the audience, the better the training will be received. Educational level is important to consider because you do not want the information to be over their heads, and you also do not want to talk down to your participants. If your audience has a mixed educational background, you need to speak to the lowest educational level.

What kind of training do they need?

Would it be best to have informal training where you have problem solving sessions, on-the-job training, one on one training, or formal training with a test? There are many different training options depending on what is required to do the jobs more safely. Be original when considering what type of training to conduct.

What must the participant learn in order to perform the job safely and effectively?

In some cases, the participants may need to know general information (e.g they have access to the written Hazard Communication program). In other cases, they may need to know more detailed information (e.g. what steps to take when the fire alarm is activated). Always know what topics should be covered and be sure to exclude any information that does not pertain to participant's jobs. For example, if you are giving a talk about saw safety, don't spend fifteen minutes talking about a band saw when the employees do not use one.

- **What should the participants expect to learn/gain from this training?**

  A good way to determine this is by establishing training goals. I like to tell people what they will learn so they know what to expect in the remainder of the training. (The next section of Chapter 3 will discuss establishing goals and objectives.)

- **Did some event prompt a request for training?**

  Was there a serious accident, or is training being held to fulfill annual retraining requirements? The audience may be in a totally different frame of mind depending on what prompted the training request. You, as the trainer, should know why you were asked to do the training, because it will affect how your present the information.

- **Is the training high or low priority?**

  Low priority means that you have not done the training in a while and it is probably time for some refresher training, even though OSHA does not have an annual training requirement. An example of high priority training is when you have new equipment arriving soon, and people need to be trained on how to use it safely before it arrives. High priority could also mean that you have an upcoming OSHA inspection.

In conclusion, a needs assessment can be a valuable piece of the training process. Without a needs assessment, you may waste your valuable time creating a training program that isn't needed. Furthermore, you may waste people's time by presenting information they do not need to know or omitting important information that they do need to know. Learning all the information from a needs assessment can change the way that you create and deliver your training. Always take the time to complete a needs assessment.

Types of Needs Assessments

One of the misconceptions about needs assessments, and a reason why people don't do them, is that they think it has to be a formal and elaborate process that takes a great deal of time. This is simply not true.

It all depends on how well you know your audience beforehand. If you are familiar with the department that made the request and you are comfortable with the tasks the employees perform, then the needs assessment can be brief. Since you already know a lot about the group and their jobs, you might only need to know the reason for the training request and the type of training that would be best suited for them.

If, on the other hand, you are not familiar with the department or the employees' responsibilities, then a more detailed needs assessment would be in order. A more detailed assessment would involve more questions for the supervisor and the employees.

In either case, the needs assessment can be formal or informal. A formal needs assessments means that you have a list of questions that you ask management and employees. It usually involves visiting the workplace before the training to actually see what the employees' jobs involve.

I prefer informal needs assessments. Most times I am familiar with the group, their job responsibilities, and where they work, so an informal conversation with the manager or supervisor over the phone is often sufficient. I also try to have a conversation with an employee. Their perception of their needs is often different than management's. Most managers are happy to give me the name of someone to talk to so I can get the employee's perspective.

Getting Management to Support a Needs Assessment

I have found management is usually very cooperative with a needs assessment. There can be times, however, when they feel it is a waste of time. There are several approaches you can use. My favorite is the old adage that time is money. If the manager is going to take the time to send the people to training, it had better be effective training, right? In order for me to make it effective, I explain that I need to ask some questions before I create and present the training.

I have also used the "buying a new car" example. Most people spend a great deal of time doing research and going on test drives before they actually buy a car. I explain to them that doing a needs assessment is my test drive before the training. I need to know everything I possibly can about the audience (car) before I give the training (buy the car).

Training Needs Indicators

Look for the following changes as they may indicate a need for training.

Changes in Personnel

When new people are assigned to a job, whether it is a new hire or a transfer, comprehensive safety training must be given to the new employee. Even if an employee is familiar with the job based on their experiences at another company, your company's safety information may be different. Never assume people will learn safety information on their own or from other employees.

Changes in Procedures

Hazards change as procedures change. Any change in the type of hazard that an employee will face will warrant training. For example, let's think about chemical pipetting. To encourage people to change the way they pipette chemicals, you would probably want to discuss the hazards of pipetting by mouth. You would also want to demonstrate how to safely use a bulb to pipette. Posting a memo or notice is not always the best way to introduce a change in procedure. Training is much more effective because it allows people to ask questions and even practice the new procedure.

High Sick Leave Rates

This may indicate a high number of injuries or illnesses on the job. See if there are any trends in the types of injuries or illnesses. If there are, then you can conduct awareness training. For example, if you notice that employees are missing a couple of months of work due to back injuries, it might be a good idea to provide back injury prevention training.

Frequent Accidents

This is definitely an indication that safety training should be conducted. Again, try to find what types of accidents are occurring and then provide the proper training. For example, if you work in a hospital and you notice that there is an increasing number of needle sticks, bloodborne pathogen training would be appropriate. You would want to include a demonstration on how to safely use and dispose of needles.

Record of High Waste

This is sometimes a problem with contaminated materials such as blood/body fluids or radiation. For example, you may question people and determine that they are throwing away uncontaminated material because it is easier than trying to decide what is and is not contaminated. They may not even know what constitutes contamination. An increase in waste may indicate that it's time for some refresher training.

Training could include what types of items are contaminated and how to safely dispose of them. You might also want to explain how to dispose of uncontaminated waste.

Poor Written and Oral Communications

People need to know what is expected of them, specifically, what safety procedures they need to follow in their jobs. If this is not communicated properly there will be a lot of confusion. Make sure that the written procedures are clear, and that supervisors are able to verbally communicate safety procedures. Refresher training helps to ensure that employees have read and understood the safety procedures.

Needs Assessment Survey

What do you think about needs assessment? Fill out the following questions to outline your procedures for needs assessment. Sometimes just writing your ideas down can help you create new ideas for improving your existing needs assessment practices.

Do you perform needs assessments before you conduct safety training?

If so, are they formal or informal?

Do you share the results of the needs assessment with management or with the employees' supervisor?

How do you conduct a needs assessment?

What have been your successes and challenges in doing needs assessments?

Overall, what do you honestly think about needs assessments?

Needs Assessment Summary

As you can see, needs assessments can be very valuable tools. At the very least, they help you to be better prepared to design and deliver the training. At best, they help you to create the most effective training possible to ensure that people receive the most pertinent information in their training.

Remember these key points about needs assessments:

- Determine as much information as you can about the audience, the type of training, and the reason for training
- Conduct either formal or informal assessments depending on your knowledge of the group for which you are providing the training
- Obtain management support
- Be aware of various training needs indicators

Goals and Objectives

After you have completed a needs assessment, the next step in designing effective training is to create goals and objectives.

Questions to Think About

▶ What should the training accomplish? What should the participants know or be able to do at the end of the training?

It is important to establish this goal before you start the training. If you are not in touch with what you want the training to accomplish, then you could miss your mark when developing the material. You are investing your time and the employees are investing their time; you should know what the training is meant to achieve.

There are many different things that the training can accomplish. For example, do you want the participants to understand concepts (e.g. how lock out tag out works), do you want them to know information (e.g. report accidents promptly), or do you want them to be able to perform detailed instructions (e.g. the emergency procedures in the event of a tornado)? There are big differences in how you would approach these three examples. Understanding what you want the participants to be able to do at the end of the training is critical.

▶ What is the nature of the objective? Do you expect evaluation, synthesis, analysis, application, comprehension, or knowledge upon completion?

This idea is explained in more detail on the following pages. You need to know how detailed the objective will be. Do you want them to be able to interpret data, illustrate ideas, explain the procedures, or know how to report something? Objectives can be more advanced or less advanced depending on the needs of the audience. More advanced objectives include evaluation, synthesis, or analysis, while less advanced objectives include application, comprehension, and knowledge.

▶ **What is expected of you, and what is expected of the participants? What do the organization and management expect?**

If you know the expectations before the training begins, there is a greater chance that you will meet those expectations. I start training with some ground rules so that everyone knows their role. The participants can expect that I will provide correct information, that I will listen and answer any questions they may have, and I will report any serious issues that may arise during the training. I expect that the participants will be open-minded, participate in any activities to the best of their ability, not interrupt when someone else is speaking, and respect the views of others.

Questions for Establishing Goals and Objectives

- What should the training accomplish?
- What should the participants know at the end of the training?
- What is the nature of the objective?
- Do you expect evaluation, synthesis, analysis, application, comprehension, or knowledge?
- What is expected of the trainer?
- What is expected of the participants?
- What does the organization expect?
- What does management expect?

Types of Objectives

There are different types of objectives that dictate the degree of learning. In this table, objectives are classified as higher and lower order objectives. Higher order objectives represent more advanced degrees of learning, while lower order means earlier stages of learning.

	Objective Type	Description	Verb Examples
Higher Order	Evaluation	To be able to determine the value of something	Interpret, solve, justify, appraise
	Synthesis	To create a new idea or concept based on a combination of various parts	Modify, reorganize, compose
	Analysis	To look at the whole picture and break it down into smaller, more understandable parts	Diagram, illustrate, differentiate, infer
Lower Order	Application	To use concepts, information, or ideas and apply them to a specific situation	Demonstrate, produce, use
	Comprehension	To understand the meaning of learned information	Explain, describe
	Knowledge	To remember previously learned information or concepts	Define, identify, match, select

Source:Rebecca Bullard, Mary Jean Brewer, Nancy Gaubas,Angie Gibson, Kathy Hyland, and Eileen Sample. *The Occasional Trainer's Handbook*; Englewood Cliffs, NJ; Educational Technology Publications; 1994, (2–9)

You can expect lower order objectives such as application, comprehension, and knowledge to require less training time, while higher order objectives such as evaluation, synthesis, and analysis demand more time. Think about what degree of learning you want, then choose the appropriate type of objective and appropriate verb.

Helpful Verbs for Writing Goals and Objectives

When writing objectives, try to use descriptive verbs that clearly define what you want the participant to be able to do at the completion of the training. The following are some possible verbs you could use.

appraise	identify	name
appreciate	illustrate	order
classify	improve	place
compare	increase	produce
compose	infer	quote
contrast	interpret	reorganize
define	know	select
demonstrate	know how	solve
describe	know relevant ways	state
develop	learn	translate
diagram	list	understand
differentiate between	match	use
display	identify	write
explain	modify	

Creating Goals and Objectives for Training

Following is a plan to create training goals and objectives. The first step is to think about the problem or what prompted the need for training. Refer to your needs assessment for the problem. Next, determine what specific information the group needs to solve the problem. Next, establish the training goal or the purpose of the training. Finally, write your objectives using descriptive verbs. Following is an example of how this would work.

Example •

Plan for Creating Training Goals and Objectives

Problem

Managers are responsible for their employees' safety training. However, the managers are not familiar with the various safety training programs that are available, which employees need training, or how often training should be conducted.

Need

Managers need to learn what safety training is available, how often it should be taught and which classes are required for the tasks that their employees perform.

Training Goal

To improve the manager's knowledge of the availability of different types of safety training so they can educate their employees on how to perform their jobs more safely.

Objectives

At the end of the training session, the managers attending this training will be able to:

1. **Identify** the type of training that is required based on the employee's job tasks.
2. **Explain** how frequently the different training sessions must be conducted.
3. **Justify** the need for training.

Note the objectives in this example are a combination of higher order (justify) and lower order objectives (identify and explain).

Exercise •

Exercise in Creating Goals and Objectives for Training

Now it's your turn. Think of some training that you will be delivering in the future or a program that you have already presented, and complete the following information as modeled on the previous page. Be sure to use active verbs for your objectives. The more descriptive the objectives, the easier it will be for you and the participants to meet those objectives.

Problem:

Need:

Training Goal:

Objectives:

Establishing Goals and Objectives Summary

Establishing goals and objectives is an important second step in designing effective training. Without clearly defined goals, you increase the possibility that you will not design or deliver what your audience really needs. You may plan to provide certain information, but your message may stray if you don't have clearly defined objectives to keep you on track. Also, most people in the audience like to know the training objectives so they will know what is expected of them.

Remember these key points about establishing goals and objectives:

- Know what is expected of you and the participants
- Make sure the objectives are clear. The more descriptive you are the better.
- Choose higher order or lower order objectives depending on the degree of learning.
- Always be aware of what the training should accomplish and what specific information the employees should learn.

Gathering Information

After you have completed a needs assessment and created goals and objectives, the next step in designing effective training is to gather information.

Where to Gather Information

It will be relatively easy to gather the appropriate information if you are familiar with the subject matter. If you are not familiar with the subject, there are a couple of different places you can turn to for help. I would suggest talking to people, reading books and journals, watching videos, attending conferences and classes, reading job descriptions and production control documentation, and searching the Internet.

People

Talk to people informally before you give the training, and even ask them to show you around their workplace. Ask what their main responsibilities are and what equipment they use most often. People are proud of their jobs and are usually happy to tell you what they know.

Books

There are many helpful books that can teach you more about the topic that you will be teaching. The American Society of Safety Engineers (ASSE), the National Safety Council (NSC), and the Occupational Safety and Health Administration (OSHA) all have excellent publications to help you. It is always a good idea to establish your own library of materials that you find helpful.

Journals

Journals are helpful because they provide the most recent information on what is happening in the safety field. They also contain articles about other safety training programs that may benefit you and your company. It is always beneficial to learn from others.

Videos

Even though you might not plan to use videos for training, they can teach you information that you might not already know. Most of us review training videos frequently; that is a great opportunity to expand your safety knowledge.

Conferences and Classes

Although conferences and classes are not always practical because of monetary constraints, they are an excellent way to gather information for future training, or even to get ideas for new topics. It's always fascinating to see what other people are doing.

Job Descriptions

Many organizations maintain job descriptions as the official documentation for each job. These documents can run the entire gamut of accuracy, although they are sometimes outdated. They may, however, contain skill and knowledge definitions, the level of accountability each job has, and hazardous or physical conditions associated with the job. Learning more about what the people actually do through their job descriptions will certainly help you to personalize the training.

Production Control Documentation

Manufacturing companies maintain detailed documentation showing how their product moves from raw materials to finished product. This documentation will also describe the locations where hazardous materials are used and in what capacity they are used. This can be very helpful information when preparing your training.

The Internet

The Internet is thriving, with more web sites appearing daily. Following is a list of helpful government and organization web sites for safety information.

Government and Organization Internet Safety Resources

Organization	Internet Address (URL)
American Chemical Society Chemical Health and Safety (CHAS) Division	dchas.cehs.siu.edu
American Industrial Hygiene Association (AIHA)	www.aiha.org
American National Standards Institute (ANSI)	www.ansi.org
American Society of Safety Engineers (ASSE)	www.asse.org
Bureau of Labor Statistics (BLS)	stats.bls.gov
Centers for Disease Control and Prevention (CDC)	www.cdc.gov
Code of Federal Regulations (CFR)	www.access.gpo.gov/nara/cfr
Consumer Product Safety Commission (CPSC)	www.cpsc.gov
Department of Energy Office of Environmental Safety and Health (DOE)	nattie.eh.doe.gov
Environmental Protection Agency (EPA)	www.epa.gov
Ergonomics Society	www.ergonomics.org.uk
National Fire Protection Association (NFPA)	www.nfpa.org
National Institute for Occupational Safety and Health (NIOSH)	www.cdc.gov/niosh
National Institute of Health (NIH)	www.nih.gov
Nuclear Regulatory Commission	www.nrc.gov
Occupational Safety and Health Administration (OSHA)	www.osha.gov

Selecting a Variety of Material

To relate to different people in the audience, you should use different types of material. This also makes it more interesting for you, the presenter. There are many different types from which to choose. Here are a few ideas:

Quote other people whose opinion the audience will value

Be sure to make a point with the quotation, and always give credit where credit is due. Do not drop names, since that can be distracting. You do not want your class trying to determine who it is you are talking about rather than listening to your point. For example, you might want to say, "as John Smith over in Human Resources always says, people are our best resource."

Share personal experiences through stories

Stories make the training more personal. It is also a clever way to encourage people to listen. However, you want to make sure that there is a purpose to your story. For example, do not interrupt your training with a story about the trials and tribulations of your child learning to walk unless it pertains to your training in some way. Always make sure that you practice the story before you use it in training. It can be awkward when a story flops.

You can also tell stories about other situations that you investigated or witnessed. I used to tell stories about accidents that happened within the same organization. People listen when they hear stories about the people that they know. Never disclose names or departments, and do not sensationalize the story when repeating the events.

Use effective statistics to illustrate your point

Statistics can be very effective—people pay attention to them. With safety training, it sometimes makes them realize that they may be the next person to be affected. If you tell people that eighty percent of all adults experience back pain sometime in their life, people listen. Make sure that you use the statistics effectively, and that you do not just ramble off meaningless numbers. Repeat key numbers that you want participants to remember. Always use a credible source and put the information in credible terms. I find that charts and graphs are a great way to show statistics. Remember that a picture is worth a thousand words.

Gathering Information Summary

In summary, the third step in designing your training is to gather information. If you are familiar with the topic, then this may be easy for you to do. If not, there are any different resources you can use to obtain interesting training material.

Remember these key points about gathering information:

- Use a variety of sources such as people, books, journals, videos, conferences, job descriptions, production documentation, and the Internet.

- Select a variety of material to keep the audience engaged. Use quotations, stories, and statistics.

Organizing Your Training

Once you have completed the needs assessment, established goals and objectives, and gathered all pertinent information, the next step is to organize your presentation.

Questions To Think About

▶ What specific information do the participants need to know in order to accomplish each objective?

After you have written your objectives, you then have a loose outline for your training. For example, if you have the following objectives:

- Identify the type of training that is required for the employees in their area based on their job tasks
- Explain how frequently the different training sessions must be conducted
- Justify the need for training

Then you should cover all these topics in your training:

- Safety training available for employees
- How often training should be conducted
- Why training is important

▶ How can the training be organized so that it is sequential and easy to follow?

Poorly organized training can be a nightmare. Have you ever been to a training session where the instructor says, "Where are we now?" or "Have I covered that information yet?" Losing your place once in a while is acceptable, but not when it occurs more than once.

Your training should follow a clear path. The easiest way to organize your training is with an opening, body, and closing. I have also heard people describe this as, "Tell them what you are going to say, tell them, and tell them what you told them." Repetition is a great way to make sure your audience remembers your main point. More information about the opening, body, and closing will be in Chapter 5, "Delivering Effective Training."

Questions for Organizing Your Presentation

- What specific information do the participants need to know to accomplish each objective?
- How can the training be organized to make it easy to follow?

Key Ideas in Organizing Your Training

- **Organize materials in a way that is most meaningful to the trainee.**

  Keep in mind this may not be the same thing as organizing materials for someone who has a better understanding of the topic. As trainers, we often fall into the trap of organizing information in a way that makes sense to us, but we often forget we have a greater knowledge of the topic than most people in the training. Try to view the material from the trainee's point of view.

  This is often difficult for me when I teach software classes. After using the software for hundreds of hours before the training, I sometimes forget how it feels to be newly learning how to use it. Ask people frequently if they understand what you are saying.

- **Start with general concepts and move to more specific.**

  For example, you could start your training by exploring how important it is to wear personal protective equipment, but then you may want to finish your talk by explaining why. If you are doing a demonstration, always show the trainees how something works before explaining why it works. For example, you could demonstrate how to perform a lock out tag out at a person's workstation, and then explain why it protects them from a possible injury.

Using a Needs Assessment to Create Objectives and a Training Outline

Training outlines are an effective way to organize your training. You can create a training outline from your needs assessment and training objectives.

Let's say that you perform the following needs assessment for nurses who request a bloodborne pathogen training session. You determine the following information:

Audience: The audience for the training will be 20–30 nurses who are new hires to the hospital. All of them have recently graduated from college and this is their first job. Most, but not all of them, are from the United States and are familiar with OSHA.

Work environment: All of the nurses will be working with patients, although they will be in different departments within the hospital.

Training request: The nursing department at the hospital requested training. They request training when they hire large numbers of nurses at one time.

Training topics to be covered: The nursing department has requested that the following topics be covered in full detail:

- Bloodborne Pathogen Standard
- Exposure Control Plan
- Universal Precautions

Time frame: Training is a high priority and must be completed during the new employee orientation, before the nurses start working. Training must be approximately two hours in length.

Based on this needs assessment you could create the following objectives:

At the completion of the training, nurses will be able to:

1. **Be familiar with** the different components of the Bloodborne Pathogen Standard
2. **Identify** which employees are covered by the Bloodborne Pathogen Standard
3. **List** potentially infectious materials
4. **Name** the key bloodborne pathogens
5. **Explain** aspects of the hospital's exposure control plan

Training Outlines

Once you have a completed needs assessment and have established the objectives, you can create a training outline. Following is a training outline you could use for the bloodborne pathogen standard training for the new nurses described on the previous page.

A. Introduction

1. Why are you here?
2. Brief explanation of OSHA

B. **Bloodborne Pathogen Standard**
   1. 29 CFR 1910.1030
   2. History of standard
   3. Who is covered?
   4. Other potentially infectious materials
      a. Body fluids, unfixed tissue or organs, or contaminated cells or tissues
   5. What is covered by the standard?
      a. Scope and applications
      b. Definitions
      c. Exposure control
      d. Methods of compliance
      e. Special requirements for HIV and HBV
      f. Hepatitis B Vaccination and Post-Exposure Evaluation and Follow-up
      g. Communication of Hazards to Employees
      h. Record keeping
      i. Declination Form

C. **Bloodborne Pathogens**
   1. HIV and AIDS
   2. Hepatitis Viruses
   3. Other pathogens

D. **Exposure Control Plan**
   1. What must be in an exposure control plan
   2. Who must have one

E. **Compliance**
   1. Universal Precautions
   2. Engineering Controls
   3. Handling Sharps and Needles
   4. Surface Contamination
   5. Personal Protective Equipment

F. **Conclusion**

As you can see, you can create a very detailed training outline from your needs assessment and training objectives. The previous example also assumes you did research to obtain your training materials. By setting up an outline, you have made an important step in organizing your training.

Many people have a difficult time creating outlines. It is sometimes hard to outline your training before you have made your training materials. For that reason, it is acceptable to start with a rough outline, complete your training materials (i.e. slides, overheads, or computer-based training) and then finish a more detailed outline.

Many software packages allow you to view your slides in an outline format. For example, in Microsoft PowerPoint, you can create your slides in "slide view" and then see the same information in "outline view." This is a quick and easy way to create training outlines from your slides.

How can you use the outline once it is created? There are two possible ways. The first suggestion is to send it to the person who requested the training. It is always a good idea to make sure you are covering all the proper information and to make sure you are "on the same page." The second suggestion is to use it as a guide when you are actually giving the training. Periodically checking your outline can quickly remind yourself what you have covered and what information is coming up.

Presentation Patterns

An effective way to organize your training is to use presentation patterns. Using patterns is effective because it is more interesting to the audience and makes it easier for them to follow the training. The different patterns discussed in this session are problem/solution, chronological, physical location, extended metaphor or analogy, divide a quote, and divide a word.

Problem/Solution

The point of this pattern is to tell the audience what the problem is and offer a solution. For example, you could start the training by saying, "There appears to be a growing number of people injuring their hands while working with saws." You could then explain how the different injuries occurred and have a discussion session so that people can participate in the problem solving. This is a very common way to organize safety training, since many training sessions are often the result of safety problems.

Chronological

In this pattern, you discuss a series of events that have happened in the past, present and future. For example, you might say, "Our accident rates have decreased substantially over the last two years. Let's discuss this trend and the changes that we made, in the hope that the rate will continue to decrease." This pattern is most effective when discussing trends.

Physical Location

Different physical locations may have different points of interest. You may want to divide your talk according to the different areas (floors in a building, different buildings, and different plants). For example, you could say, "I'm going to divide my chemical safety talk by the different labs that we have in our department because each lab operates differently." This pattern is only effective if your participants come from different locations.

Extended Metaphor or Analogy

This is using something familiar to the audience (for instance, driving a car) and comparing it to how you will proceed through your presentation. For example, you could say, "As we talk about back injury prevention this afternoon, I will be starting the engine by talking about the anatomy of the back. Then we'll cruise the highway while discussing different injuries that occur. Finally, we'll park the car by discussing ways to prevent these accidents from occurring." This can be an effective pattern, but it is sometimes difficult to develop appropriate metaphors or analogies.

Divide a Quote

For this pattern, you take a quote that is widely used and break it down into different sections that represent the sections of your talk. For example, you could break down the following quote: "Safety is always our primary concern." Then you could continue the training with the following questions:

1. What is safety?
2. What are our concerns? As employees we have to think about production deadlines, saving time and money, being a cooperative employee, and being safe.
3. Is safety always our primary concern? Why or why not?

This is a fun way to organize your training, and people are usually able to remember your information better. This is only effective if you use a quote that people are familiar with.

Divide a Word

This pattern is fun, because you take each letter of a word and use the letters to describe your ideas. For example, you could break down the word "accidents." You could start you training by saying, "Today we are going to talk about accidents. A stands for acting in a careless manner..." Then you would continue with all of the remaining letters in the word. People generally remember acronyms, so this would be very effective in helping people to remember the key points of your training.

When using these presentation patterns, make the patterns obvious and make sure the pattern you use is appropriate. For example, if you were talking about accident trends, the most effective pattern would be chronological.

Exercise •

Creating Patterns for Training

Take a few minutes to determine examples of different presentation patterns for the following safety topics. Some suggested answers are in Appendix F.

Presentation Pattern	Type of Training	Example
Problem/Solution	Hazard Communication	
Chronological	Lock Out/Tag Out	
Physical Location	Hazard Communication	
Extended Metaphor or Analogy	Personal Protective Equipment	
Divide a Quote	Back Injury Prevention	
Divide a Word	Ergonomics	

Organizing Your Training Summary

Organizing your training is an important step in designing effective training. Lack of organization will frustrate your audience and cause you to lose credibility. Being organized also helps you to cover all the appropriate information. When organizing your training, think about creating a training outline from your needs assessment and objectives. Also consider using a presentation pattern to make your training more interesting.

Remember these key points about organizing your training:

- Organize materials in a way that is most meaningful to the trainee
- Start with general concepts and move to more specific information
- Creating outlines helps to determine if you are presenting extraneous information or have forgotten vital material
- Training outlines can be created from needs assessment and your objectives, or they can be created after you have made your materials (slides, overheads, or software presentation)
- Presentation patterns make training more interesting and increase chances that people will remember your information
- Make sure the presentation pattern you choose makes sense

Training Methods

After you have organized your presentation, the next step is to choose your training method. Possible training methods include Lecture, Discussion, Demonstration, Structured Exercise, Case Study, Role Playing, In-Basket, Games, and Brainstorming.

Questions to Think About

▶ **What training methods will help the participants to meet the training goals and objectives?**

If your objectives only hope to give the participants information, then a lecture would suffice. If you hope to have them retain the information and be able to apply it on their job, then you might want to try a demonstration or role-playing. The more difficult objectives (i.e. higher order) will require more audience participation while the easier objectives (i.e. lower order) will mostly involve you, the trainer, with little involvement from the audience.

- **Do the methods chosen allow for participant involvement?**

  Always try to involve the audience in some way—the more involvement, the better. People remember more if they are doing something. Although thinking of ways to involve the audience may be challenging, it will clearly make your training more effective.

- **Do the methods chosen offer variety?**

  Variety is important. Nobody wants to sit and listen to someone talk for hours. When I teach classes, I like to speak for a while and then follow up with group activities. In the end, I also like to play games since people leave in a great mood and they have a positive attitude about training.

- **When deciding on methods and materials, have the following issues been considered?**

  - Budget and monetary constraints
  - Availability of equipment
  - Lighting, size, and set up of the room
  - Equipment portability

  It is generally a good idea to see what restrictions you may have before you decide on a training method. Unfortunately, budget constraints can sometimes limit the type of training method that you use. You also have to consider the structure and layout of different training rooms, which can limit the training method you choose. For example, it's awkward to brainstorm when you are in a lecture hall that seats two hundred people.

Questions for Training Methods

- What training methods will help the participants to meet the training goals and objectives?
- Do the methods call for participant involvement?
- Do the methods offer variety?
- Have you considered budget and monetary constraints?
- Have you considered availability of equipment?
- Have you considered lighting, size and setup of the room?
- Have you considered equipment portability?

Selecting a Training Method

Following are some typical methods you could use:

Method	Trainer's Role	Use it When...
Lecture	Present in front of group	Large group of people needs to share the same information
Discussion	Lead discussion and keep training on track	You want to gather or pass on information or solve a problem. Also use it when the trainees have a good knowledge of the topic and they are willing to share experiences and listen to others.
Demonstration	Physically imitate proper way to perform activity and ask employees to practice	Actions speak louder than words
Structured Exercise	Facilitate exercise	You want participants to practice new skills
Case Study	Present case then have participants analyze and draw conclusions	You want to apply concepts or evaluate ideas
Role Playing	Oversee participants acting out a situation that leads to a discussion	You need to see situations from different perspectives since learners are asked to act the parts of other people
In-Basket	Create papers containing problems and ask participants to solve	You want to teach decision-making under a time constraint
Games	Be fun! Create interesting ways to convey information with games	You want to elicit audience participation and increase interest in training. This also facilitates cooperation between groups.
Brainstorming	Encourage participants to list different ideas without fearing negative ramifications.	You are having a hard time coming up with a solution to a problem

Adapted from: Robert Pater, *How to Make High-Impact Safety & Health Presentations*, 2nd ed., Des Plaines, IL; American Society of Safety Engineers, 1995 (22)

Advantages and Disadvantages of Using Different Training Methods

Lecture

- You can speak to a lot of people at one time.
- There is little or no trainee interaction and there is an absence of trainee accountability. You need to have many different visual aids.

Discussion

- Involves the audience. It also allows the trainer to learn about other people's views. Great method for problem solving and team building.
- You may have a few people who talk too much and others who are too intimidated to speak their views. The group also has to be relatively small. The trainees must have a positive attitude and be open to different views. May be time consuming.

Demonstration

- A wonderful way to show the audience exactly what you are talking about. Great way to focus on basic procedures.
- It may be expensive to get props and it may be time consuming. Trainer must be skilled and be able to explain each step being performed. You may have to limit the number of people if you are demonstrating something small.

Structured Exercise

- Good way to get people involved.
- May be time consuming, depending on level of difficulty.

Case Study

- Higher degree of learning because participants need to apply the concepts they have learned. Trainees learn to identify and solve problems while making their own decisions.
- It is sometimes difficult for the trainer to create effective case studies that relate to people's jobs.

Role Playing

- Allows people to see problems from different angles. Increases personal involvement and practical experience.
- May be intimidating for an introverted group.

In-basket

- Helps people to make decisions on many different problems related to the subject.
- People may not be happy with the problem they are asked to solve.

Games

- Non-threatening and enjoyable way to see if participants remember the training. Great for team building. Increases interest in training.
- Trainer has to create the games. The group may become rowdy. May require a discussion session held after the game to reinforce what should have been learned.

Brainstorming

- Great way for people to suggest ideas without the fear of being judged. You can exchange many different ideas in a short period of time. Stimulates thinking and generates ideas from several perspectives.
- Group must participate in this for it to be effective. People may be frustrated if their ideas aren't used. Trainees must have a background in the subject matter to be able to contribute ideas.

Exercise •

Training Exercise: Selecting a Training Method

Take a few minutes to determine the most effective way(s) of conducting the training for the following safety training classes. Some suggested answers are found in Appendix G.

Safety Training	Possible Training Methods
Accident Investigation	
Back Injury Prevention	
Lab Safety	
Lock Out/Tag Out	
New Employee Orientation	
Respiratory Protection	
First Aid	
Hearing Conservation	
Material Safety Data Sheet Training	
Confined Space	
Cardiopulmonary Resuscitation (CPR)	
Radiation Protection	
Asbestos Awareness	
Ergonomics	
Bloodborne Pathogens	

Training Methods Summary

Selecting a training method is an important step in determining the best way to meet the training goals and objectives. Nine possible methods include: Lecture, Discussion, Demonstration, Structured Exercise, Case Study, Role Playing, In-Basket, Games, and Brainstorming. There are advantages and disadvantages of using each. When choosing a training method, consider people's background, the desired level of learning, the number of trainees, and the time allotted.

It is possible to use more than one method in a training class. For example, you could start the training with a lecture, have a case study to apply the concepts, and then conclude the training with a brainstorming session.

Remember these key points when selecting your training method:

- When possible, choose a training method that involves audience participation.
- Training methods should help trainees to meet the higher order and lower order objectives you have established.
- Consider availability and portability of equipment, the set up of the room, and your budget.
- Size of the audience or training room may limit what training methods are available.
- It is possible to use a combination of training methods.
- Be creative.

Audio-Visual Aids

Once you have established your training method, you need to select the visual aids you will use to convey the information. Visual aids discussed in this chapter include videotapes, handouts, overhead projector and transparencies, laptop computer and LCD projector, flip chart, slides, actual object, and computer-based training.

General Tips for All Types

Never put too much information on a visual

You want the participants to spend their time listening to you, not reading a slide. Remember the KIS principle: Keep it simple! A good rule for slides is to have no more than seven lines with seven words on each line. See page 61 for an example of a good and bad visual.

- **Always make it pleasing to the eye**

  There is nothing worse than a sloppy visual. If you want people to pay attention, you have to impress them. Use soothing colors such as greens, blues, and purples rather than upsetting colors like reds, oranges, and yellows.

- **Do not let the audio-visual materials interfere with your presentation**

  Remember that audio-visuals support what you are saying without distracting the audience. A good example of materials interfering with your presentation would be the different builds and transitions that people use on PowerPoint presentations. You can spend more time trying to determine what the next build will be rather than concentrating on what is being said.

- **Never leave a "blank page"**

  This tends to make people uncomfortable so make sure that you always have something on the screen other than a glaring white light. Have the participants look at the present slide while you finish talking. If you are pressed for time you might want to flash the next slide to encourage them to wrap up their question.

- **Know your visual**

  The time to figure out what to say is not when you show the visual. Always practice BEFORE your presentation and have notes if you need them.

- **Know the order**

  This will aid in having smooth transitions. Printing your outline is a good way to remember the order. With an outline you can see your progress through the presentation, and you will always know which topics you have covered and which ones are coming up.

Example of a Good and Bad Visual

What is the Chemical Hygiene Plan?

- 29 CRF 1910.1450
- Way for lab workers to find out about hazardous chemicals

Good Visual

- Clear
- Easy to read
- Title in sans serif font
- Text in serif font
- Simple bullets
- Appropriate graphic

What exactly is the OSHA Chemical Hygiene Plan (CHP)?

- OSHA regulation can be found at 29 CFR 1910.1450
- The Chemical Hygiene Plan standard was created in 1991 by federal OSHA
- Way for lab workers to find out about hazardous chemicals
- What is a hazardous chemical?
- - any substance which is toxic, corrosive, an irritant, a strong sensitizer, is flammable or combustible or generates pressure through decomposition, heat or other means, if such substance may cause substantial

Bad Visual

- Too much information
- Too many different fonts
- No graphic
- Too wordy
- Inappropriate bullets
- Very distracting

Using Different Types of Media

Here are some suggestions for different types of media:

Videotapes

Make sure that the video will relate to your audience. Do not show a manufacturing video to a academic audience. They will not be able to relate to the different situations.

- Do not show a video that takes more than half of your training time. Videos are there to support your training, not replace it.
- Cue the video beforehand. Showing the credits before the training can be a waste of time since the participants should only see the pertinent material. You can also fast-forward through sections that you don't feel are important—but only do this once, since people may feel cheated that they are not seeing the whole video.

- Use videos to show things that you cannot demonstrate in a classroom setting. For example, I would not want to demonstrate what happens to a pyrophoric chemical when there is a spark nearby.
- Make sure that the screen is large enough to view and that the sound is loud enough. Ask the audience if you are not sure. Make sure that you can clearly see the screen from all available seats. It is a good idea to run the tape before your class arrives. Move around the classroom to check the audio and visual from different perspectives.

Handouts

- Make sure that they are legible and clean. Nobody likes to read handouts with smudged fonts or crooked words.
- Staple or clip your handouts together. This will decrease the possibility that people will lose the material.
- Number the pages. If your handouts are coming from different sources, write the page numbers in yourself.
- Have copies for everyone present at the training. This usually means bringing extra handouts. If you run out, make sure you send copies to the people who need them as soon as possible.
- Pass handouts out at the appropriate times—prior to, during, or after the training. If you pass them out at the wrong time, people will spend their time reading the handouts and not listening to the training.
- Leave lots of white space for people to write their own ideas or take additional notes.

Overhead Projector and Transparencies

- ALWAYS have an extra bulb—you never know when the one you are using will burn out.
- Make sure that your transparencies fit the screen and that words are large enough. Most overhead screens are set up to view pages that are formatted as "portrait," so do not set up your pages "landscape."
- Speak louder to compensate for the noise of projector.
- Position yourself so you do not block the view of the screen. This can be a practical problem in a small room that is filled to capacity.

Laptop Computer and LCD Projector

- Get the software booted before your audience arrives. Do not spent time making adjustments to your slides as the people arrive.

- Make sure you know what version of presentation software is on the laptop and save your file in the same version.
- Practice the order of the slides and be able to easily move forward and backward.
- If the laptop is operating from a battery, make sure it's charged before the presentation.
- Do not use a lot of distracting transitions and builds as you move from one slide to another and as your points are added to the screen.
- Test the presentation on the laptop and LCD projector that you are going to use. Colors look different from one computer to another. The LCD projector may also cut off some of the screen, so make sure that you have an adequate border.
- Have hard copies of the slides or overheads ready in case you cannot get the equipment to work.
- If you need to connect to the Internet, test your connections before the training begins.

Flip Chart

- Have plenty of pens to write and make sure they have ink.
- Only use with small groups.
- Write LARGE (three-inch lettering) and use dark ink colors.
- Watch your spelling as you write. Misspelled words can be distracting.

Slides and Slide Projector

- Make sure the room is dark enough to read slides, but not so dark that people take a nap.
- Have a white background on your slides so that it provides enough light in the room.
- Check the orientation of the slides to make sure they are not accidentally backwards or upside down.

Actual Object

- Point out important aspects of the object while demonstrating.
- Allow the participants a lot of time to handle the object and/or practice with it.
- If the object is fragile or expensive, tell them so they can take extra care in handling it.

Computer-Based Training

- Make sure that the training is specific and not a generic package. There are many companies that can create customized packages for your company.
- Make sure that the availability and portability of the computer is not a concern.
- Test your employees' comfort level with computers. You may have employees who never use computers. Safety training is not the time to introduce them to the marvels of the computer age.
- If needed, provide clear written directions on how to enter and exit programs. Also, give directions on what to do if error messages occur during the training, and who to contact for technical support.
- Always have a person available to answer questions. Computer-based training can be effective, but you need to always provide a human voice for people to contact if they have any questions about the training.

Exercise •

Pros and Cons of Different Types of Media

Take a few minutes to write the pros and cons of the different types of media. See Appendix H for some possible answers.

Type of Media	Pros	Cons
Videotapes		
Handouts		
Overhead projector and transparencies		
Laptop computer and LCD projector		
Flip chart		
Slides and slide projector		
Actual object		
Computer based training		

Ranking Training Aids on Seven Dimensions

Training Aid	Trainee activity required	Trainer activity required	Number of senses used	Simplicity in production	Simplicity in use	Cost to produce training aids	Cost to use training aids	Portability
Videotape	5	4	4	8	4	2	7	2
Handouts	4	5	8	3	2	7	6	1
Overhead projector & transparencies	6	3	5	5	5	4	3	6
Laptop computer & LCD projector	7	2	6	4	8	6	2	7
Flip chart (trainees write)	2	6	3	2	3	8	5	5
Slides & slide projector	8	1	7	6	6	3	4	4
Actual object	3	7	2	1	1	5	8	3
Computer based training	1	8	1	7	7	1	1	8

The six separate items have been ranked from 1 (highest) to 8 (lowest) based on seven different criteria. Results are based on the opinion of the author as a result of training experiences.

Visual Aids Summary

There are many different visual aids; the following were discussed: videotapes, handouts, overhead projector and transparencies, laptop computer/LCD projector, flip charts, slides, actual objects, and computer based training.

Remember these key points when selecting visual aids:

- Consider the advantages and disadvantages of using different aids
- Consider size of audience and constraints in the room (e.g. you may not be able to close the blinds)
- Be prepared for your equipment to fail. Always have backup transparencies, slides or handouts

Remember these key points when using visual aids:

- Never leave a blank page
- Be familiar with all the information on the audio-visual
- Know the order of slides or overheads

Testing and Developing Questions

We are now approaching the end of our chapter on designing effective training. We have discussed needs assessment, training goals and objectives, gathering information, organizing your training, training methods, and audio-visual aids. The next step in designing effective training covers different testing mediums and ways to develop questions.

Questions to Think About

- **What is the best way to determine if the audience has learned what you want them to learn?**

There are different ways to see if people have learned all the information from the training objectives. Tests are effective but you could also use an informal conversation at the end of class. If you use a test you should think about the different types of tests that you could use.

- **What are some different ways to test or quiz?**

Most of us think of formal tests where an instructor passes out a series of questions on a piece of paper. For some types of safety training, this is important for record keeping purposes, and it ensures that the participants have an understanding of the material. If a formal test is not required, then discussions are effective. Discussions are good only if you want to get a feel for the level of learning. Remember to include everyone.

One of my favorite ways to give tests is through crossword puzzles. It's an interesting way to make sure that people understand the material. I like to have them work on the puzzle in teams, but you can also have them work individually. People are so busy trying to solve all the puzzles that they forget they are being tested. Appendix G has some examples of crossword puzzles for chemical safety, laser safety, and hearing conservation. Software applications for creating crossword puzzles and other word games are available at most office supply stores.

Tests may also cause anxiety. It is very important to warn people at the beginning of training that they will be tested. You must be very clear regarding the nature and use of tests. Will they be formal or informal?

Will the score be recorded? Will my supervisor receive my grade? Is passing the test a condition of my continued employment?

There will be situations where people will be tested on the information and required to achieve a minimum score. In this case, you need to make sure that the questions are clear and were adequately covered during the training.

- **How do you decide what questions to use?**

  This is a tough question. You have to determine your most important points and then create questions that test people's understanding of these points. Do not test people on miniscule facts in an effort to make sure they paid attention. Do not create trick questions that will confuse the class and reduce the effectiveness of your training.

Questions for Testing and Developing Questions

- What is the best way to determine if the audience has learned what you want them to learn?
- What are some different ways to test or quiz?
- How do you decide what questions to use?
- How difficult should I make the questions?

Helpful Hints to Remember

- **Make the test relatively easy to pass**

  The purpose of the test is to show that the individual listened to the training, not that they are proficient in the subject. You should easily be able to tell if people were attentive, or whether they took a nap during the training.

- **Enable the participant to complete the test within 10–15 minutes**

  Again, the purpose is to make sure that they have grasped most of the information. The longer the quiz, the more confused and irritated the people will be. Also, remember that most people do not want to be at the training (especially if it is mandatory), and they will need to return to their job responsibilities quickly.

- **Always go over all of the answers at the completion of the quiz**

  You always want to make sure that everyone understands what they got right and wrong. If they got a question wrong, always explain WHY it is wrong and then tell them the correct answer. Do not just simply say it's wrong. Correcting a test can be your last chance to impart the desired information.

- **Make the testing medium interesting if possible**

  Try different ways to test the participants. For example, instead of having a multiple choice or true/false test, give them a crossword puzzle on the subject. If you have a large group you can also try a short game of safety jeopardy with teams. Safety jeopardy is similar to the jeopardy game on television, except that the questions relate to safety. See Appendix E for a sample of safety jeopardy.

Guidelines for Written Tests

To measure:	Consider test type:	Advantages	Disadvantages
Memory	True/False	• Easy to score • Instructions are easy to understand	• Must have clear statements • Encourages guessing • Students tend to think statements may be trick questions
Knowledge of facts or ability to discriminate between possible alternatives	Multiple choice	• Easy to score • Guessing is not encouraged	• Difficult to construct • Not adapted to complex answers
Knowledge of facts (labeling)	Identification	• Easy to construct and score	• Obtaining drawings, photographs, and actual equipment takes considerable time
Discrimination between related information	Matching	• Easy to construct and score	• Tendency for cues to correct answers

Adapted from: Rebecca Bullard, Mary Jean Brewer, Nancy Gaubas, Angie Gibson, Kathy Hyland, and Eileen Sample, *The Occasional Trainer's Handbook*; Englewood Cliffs, NJ; Educational Technology Publications; 1994 (5–25)

Examples of Well-Constructed And Poorly-Constructed Test Questions

True/False

Well-constructed question:

When a chemical exposure has occurred in your eyes, use an eyewash for fifteen minutes, then seek medical attention.

This is a well-constructed question because it is clear and it reinforces one of the key points of the training. It is also very specific because it states the length of time to use the eyewash.

Poorly constructed question:

If you happen to spill something in your eye while you are working with chemicals, any chemicals, it is usually a good idea to use an eyewash for a while then go to the emergency room or your personal physician, whichever is easier for you.

This is poorly constructed because it is too wordy, not direct, and not specific enough.

Multiple Choice

Well-constructed question:

One hazard that may cause someone to slip is:

a. dry floor
b. floor with grease, oil, or water
c. hardwood floor
d. clean floor

This is well constructed because it asks the person to choose only one option. Although you have to read all the options, the answer is quite easy to determine.

Poorly constructed question:

Which combination of floor conditions is likely, under normal conditions, to cause someone to slip or fall while they are working during the day or night?

a. dry or clean floors
b. clean floor or hard wood floor
c. floor that has grease, oil, or water spilled on it
d. none of the above

This is a poorly constructed question for the following reasons:

- too wordy
- the possible answers are too confusing and tricky
- not a good idea to use "none of the above"

Also be careful giving answers like "a and b, but not c" or "everything except a" because you are asking people to evaluate the answers as they are related to each other, rather than having them pick the correct answer.

Testing and Developing Questions Summary

Developing questions for tests and quizzes can be difficult, but tests can be a valuable way to determine if the trainees understood the information you presented. Although most employees dislike taking tests and quizzes, they are sometimes required for mandatory training. Make sure the trainees know how the test results will be used, or how the results will affect them or their job. Also consider different testing mediums.

Remember these key points when developing test questions:

- Don't ask trick questions, be direct
- Test only the key points
- Make the questions short and easy to answer
- Be specific to avoid confusion
- If using a written test consider True/False, Multiple Choice, Identification, or Matching

Logistical Arrangements

The next step in designing effective training discusses logistical arrangements such as the facilities, room arrangements, and equipment.

Questions to Think About

- **Are you familiar with the facility in which you are training?**

  This does not only mean that you know how to get to the building or room. It also means that you can locate the emergency exits, restrooms, water fountain, vending machines, telephones, and technical help if you need it. Also make sure you can direct people where to park and can give permits or passes for their vehicles.

- **Is the room arranged in a way that is conducive to learning?**

  The way that the room is laid out logistically can make a big difference, depending on the type of training you are doing. My favorite is the U-shape. It allows the participants to see the instructor and each other so that you can easily transition from lecture to group activities. Different seating arrangements will be discussed later in this section.

- **Can you accommodate participants with special needs?**

  Special needs can mean anything from wheelchair access to seating in the front for people who have bad eyesight or hearing. It is usually helpful to know what special needs participants may have, so you can make individuals comfortable without embarrassing them.

- **Is the equipment set up for convenient use without interfering with the participants' ability to see?**

  Generally speaking, you want to have your equipment set up in the front of the room. Make sure that you know where you are going to stand. Also, make sure that the equipment or electrical cords are not in the participants' way. I have had to set up slide projectors in the fifth row of an auditorium just to make sure that the image is large enough. Luckily, most people sat in the back seats so the projector and cords were not in their way.

Facilities, Room Arrangements, and Equipment

Logistical arrangements refer to the classroom facilities, the way that you have your room arranged and equipment set up. Following are some things you want to remember when planning your class. Most of these tasks should be done before your training, especially if it is a new type of training or a new location.

Facilities

- Check schedules, flyers, and announcements to make sure that they have the correct date and time. My favorite mistake is when people accidentally advertise classes that are at 1 a.m. instead of 1p.m.
- Determine seating arrangements and room size. Possible seating arrangements can be found on page 74.
- Be familiar with the location of the emergency exits, restrooms, water fountains, vending machines, and telephone.
- Make sure you have someone to contact if you have problems with the facilities. You may need to adjust the heating and cooling for the room as the day progresses.
- Know what parking is available.

Room Arrangements

- Arrange the room according to your style and the appropriateness of training.
- Test acoustics and obtain a microphone if you need one.
- Notice the location of heating and air conditioning vents, as they may cause noise or physical discomfort.
- Locate the electrical outlets (2 or 3 prong) and the light switches.
- Check the windows for blinds or curtains if you plan to use audio-visuals.
- Test the room lighting to see if you can dim the lights.
- Make sure that you can accommodate participants with special needs.

Equipment Set-up

- Test all equipment prior to the beginning of each class.
- Have backup supplies (pens, paper, projector bulbs, extension cords, 3-prong adapter) ready to use.
- Have extra copies of training materials in case extra people show up.

Seating Arrangements

Classroom Style

- Any audience size
- Allows easy use of training materials
- Allows adequate room for movement
- Effective if trainees need to write

Round Table

- Medium to large audience
- Allow 54 inches between seats
- Used for meal functions, study groups, or discussion groups
- Difficult for some participants to see the visuals

Theatre Style

- Medium to large audience
- Uses only chairs, no tables
- Difficult for people to use a manual or write
- May require the use of a microphone for large rooms
- Does not encourage group interaction

U-Shaped

- Small or medium audience
- Tables are shaped in a U with the trainer at the closed or open end
- Provides a clear view of any demonstrations
- Encourages group interaction

Hollow Square

- Small to medium audience
- Table forms a square with the seating on the outside
- Encourages participation and exchange of views

Conference Style

- Small audience
- Only one table is used with the people facing the center of the table
- Effective for conference meetings and discussion groups

Logistical Arrangements Summary

Considering your logistical arrangements before your training can help you be prepared. Although we often forget logistical details, they can make a big difference between a pleasant or unpleasant training experience. It will be difficult for you or your trainees to learn if they are not physically comfortable.

Remember these key points when considering logistical arrangements:

- Make yourself comfortable with the facility you are training in. Know where to find the emergency exits, restrooms, water fountains, vending machines, telephones, and technical help.
- Arrange the room so it is comfortable and conducive to learning. Possible arrangements include Classroom Style, Round Tables, Theater Style, U-Shaped, Hollow Square, and Conference Style.
- Be able to accommodate people with special needs.
- Set up equipment so it does not interfere with your ability to move around freely or people's ability to see.

Evaluations

The last step in designing effective training is to ask the audience to evaluate you and the training.

Questions to Think About

▶ **When doing an evaluation, what outcome are you trying to measure?**

There are many different outcomes that you can measure. It is important to recognize what you are measuring BEFORE you create the evaluation. Most of the time I am trying to determine two things: whether the material is well-written and whether I did a good job teaching the class. You may also want to know if it was a convenient time for training and if the location was chosen well.

▶ **What did the participants consider to be beneficial/not beneficial in the training?**

This is very helpful. What you think is beneficial may not be. Once you find out what is not beneficial, then you can either remove portions of the training or just spend more time on what they thought was helpful.

▶ **Were the training objectives met and if so, how do you know?**

This is important because you want to make certain you covered everything. It can be hard to tell whether you met the objectives or not. You can ask the participants in the evaluation, but there is the possibility that the respondent is not focused on the class objectives, may be in a rush to leave, or may not remember the objectives. Testing is usually a good way to tell if the objectives were met. If they complete the test successfully then you know that you probably covered all of your objectives.

- **Has on-the-job performance changed as a result of the training?**

This requires following up with the participants or management after the training. This is an important part of the training process but something that is often overlooked or ignored. All it takes is an additional evaluation form, usually much shorter, that you send to the participants a month after the training. For example, if you conducted training on personal protective equipment (PPE), wouldn't it be good to know if people were more likely to wear their PPE now that they have had the training? If this is not the case, then it is possible your training was not as effective as you initially thought.

Why Do Evaluations?

You perform evaluations to see if the participants were properly trained.

You may also want to:

- See if the objectives were adequately met
- Evaluate the reasons for the success or failure of the training
- Receive feedback from the participants
- Judge the effectiveness of the training
- Learn how the training was received to improve the presentation
- Determine if on-the-job performance has changed as a result of training
- Evaluate your delivery of the training

Remember, you are doing an evaluation to measure the outcome against the expected results.

Creating Effective Evaluations

There are five steps you should take when creating an evaluation for the participants to complete:

1. Determine what outcome you want to measure.

For example, you may want to know if the trainer was effective in his/her delivery, or if the training environment was too hot or too cold. Was it the proper time of day for training and was the lighting adequate? Did the participants think the training was valuable? What additional training would the participants like to have?

Finding out what outcome you are measuring helps you to focus on the questions you ask in your evaluation.

2. Decide what type of evaluation to use.

There are various types that are used. For example, you can ask for short responses to questions or have a graded scale. The graded scale could have numbers from 1 to 5 (1 being poor and 5 being excellent). It could also have statements like strongly agree, agree, not certain, disagree, and strongly disagree. See the following pages for possible evaluations.

3. Select the method you want to use to deliver the evaluation.

Some methods include delivering the evaluation orally, in writing, or through a web page. The most popular and probably the easiest is to ask people to write their thoughts on paper. This method is effective, because it is easy for people to complete and it can be anonymous. It is also easy to process the information. Remember you can deliver your evaluation the way that you like. What is important is whether you receive the information you need to measure the outcome.

4. Determine how to tabulate the data and how to report it.

For example, are you interested in the mean, the medium or the mode for the answers? Will the results be reported to the trainer, the trainer's supervisor, or the person requesting the training? Knowing this before you create your evaluation will make tabulating the results easier.

5. Understand how the evaluation results will be used.

For example, is the purpose to decide whether the trainer will receive a merit raise or pay increase? Is it to change the environment of the training or to decide how the training materials were received? Is it to make constructive changes for the next time the training is delivered? As you can see, you could have many different ways to use the evaluation results. Be aware of how the results will be used before you create your evaluations.

When instructors are given the feedback, make sure they are trained in how to interpret the information. When I do consulting for other organizations I sometimes have a difficult time interpreting their evaluations, so I always take a few minutes to have someone explain the results.

Sample •

Sample Evaluation #1

Agree or Disagree

Name of Training: ______________________	**SA = strongly agree**
	A = agree
Date of Training: ______________________	**U = uncertain**
	D = disagree
	SD = strongly disagree

The objectives of the session were clearly stated at the beginning.

SA A U D SD

Throughout the training class, I knew what was expected of me.

SA A U D SD

The presentation was clear and easy to understand.

SA A U D SD

The visual materials (handouts, overheads, and slides) were easy to understand.

SA A U D SD

I feel that this was valuable training for me to receive.

SA A U D SD

The training was held at a good time.

SA A U D SD

The training was held in an appropriate and comfortable location.

SA A U D SD

I feel that the trainer did a good job.

SA A U D SD

Sample Evaluation #2

Answer the Following Questions

Name of Training: ______________________________

Date of Training: ______________________________

Please answer the following questions to the best of your ability so that we can improve future training sessions.

What did you enjoy most about the training?

Which part of the training did you find most useful for your job?

What did you enjoy least about this training?

What part was the least useful for your job?

What would you like you have added to the training?

What would you do to improve this course?

What was the instructor's best quality?

What does the instructor need to improve upon?

Name (optional) ______________________________

If you sign your name, the instructor may contact you to clarify your comments. Your suggestions can be very helpful in improving this class.

Sample •

Sample Evaluation #3

Poor to Excellent

Name of Training: ______________________________

Date of Training: ______________________________

Please answer the following questions to the best of your ability so that we can improve future training sessions.

How would you rate the following questions with 1 being poor and 5 being excellent? Please circle your response.

	Poor				Excellent
The instructor's ability to answer questions	1	2	3	4	5
The instructor's knowledge of the subject	1	2	3	4	5
The instructor's interest in the subject	1	2	3	4	5
The instructor's preparation	1	2	3	4	5
The instructor overall	1	2	3	4	5
The attainment of class objectives	1	2	3	4	5
The length of the class	1	2	3	4	5
The class materials (handouts)	1	2	3	4	5
The class environment (room size, temperature, and lighting)	1	2	3	4	5
The overall class	1	2	3	4	5

Additional questions:

My favorite thing about the instructor:

One thing the instructor should improve:

My favorite thing about the training:

One way to improve the training:

Sample •

Sample Evaluation #4

Trainer Evaluation

Trainer:______________________________ Date: ______________________

	Poor	Fair	Good	Excellent
Prepares trainee for learning				
Provides learning objectives				
Motivates in terms of HOW the material is to be used				
Conveys enthusiasm to the trainees				
Establishes and maintains rapport in a professional manner				
Senses trainee needs				
Holds the respect of the trainees				
Allows trainees to communicate with him/her				
Provides clarification and reinforcement of the learning objectives				
Handles behavior problems in an effective manner				
Selects and uses media and/or facilities effectively				
Uses training skills effectively				
Conducts sessions on the trainees' level				
Uses good questioning techniques				
Uses a well-modulated voice				
Shows care in personal appearance				
Demonstrates flexibility in adjusting to unplanned learning situations				
Displays knowledge of subject matter				
Demonstrates adequate ability to use learning materials				
Manages time effectively				
Shows evidence of careful planning of presentation				
Evidence of interaction between trainer and trainees				
Evidence of group involvement				

Source: Douglas G. Mayo, and Philip H DuBois, *The Complete Book of Training: Theory, Principles, and Techniques*; San Diego, CA; University Associates, Inc.; 1987 (59)

Don't Forget Your Own Evaluation

As we have seen, having the participants evaluate your performance as a trainer and the training materials is very beneficial. However, we often forget to evaluate ourselves. The following are some questions you may want to ask yourself after you have completed your training. Remember that the goal is to improve the training materials and improve your training techniques. We can always learn from our mistakes.

What Happened During the Session?

- Look for overall patterns of behavior that moved things along or slowed them down
- For future use, note the things that went well
- Recognize the problems that came up and determine ways to avoid them in the future
- Brainstorm about new ideas for the next time

How Did the Training Compare with Your Original Plan?

- Do the training objectives still seem clear after presenting the material?
- Were the objectives met upon completion of the training?
- What was helpful in achieving objectives and what hindered achievement?
- Did you have enough time to present the material?
- Did people participate? Why or why not?
- What would you include or leave out the next time you presented the material?
- What new things might you try?

Evaluation Summary

As you can see, evaluations are a very effective tool in determining how your training was received. Your perception can sometimes be very different from the participants', and it is important to learn their thoughts. Constructive criticism from the audience can help you to improve your training techniques by making you aware of certain traits you may have that are distracting. Constructive criticism of the actual training can help you improve your audio-visuals and refine what information you present.

Remember these key points for evaluations:

- Determine what outcome you are measuring
- Decide what type of evaluation to use
- Consider how you want to deliver the evaluation

- Know how the data will be tabulated and reported
- Understand how the results will be used
- Make the evaluation easy to fill out
- Don't forget your own evaluation.

Checklist •

Training Design Checklist

Use this form to help design your training.

Needs Assessment	Who needs the training?	
	Who requested the training?	
	What kind of training do they need?	
	What is the educational level of the people who need the training?	
	What do they need to know?	
	What event prompted the need for training?	
Goals and Objectives	Is the training high or low priority?	
	What should the training accomplish?	
	What is expected of the trainer?	
	What is expected of the participants?	
	What does management and the organization expect?	
	Are my objectives clear?	
	Do I want higher order or lower order objectives?	
Gathering Information	Where can I look to gather information?	
	Have I selected a variety of material?	
Organizing Training	How can the training be organized to make it easy to follow?	
	Is the training organized so it is meaningful to the people?	
	Have I created an outline?	
Training Methods	What presentation pattern could I use?	
	What training method(s) could I use?	
	Do the methods call for audience participation?	
	Do the methods offer variety?	

Training Design Checklist (cont'd)

Visual Aids	What audio-visual aids will be best?	
	Are my visuals clear and easy to read?	
	Am I familiar with all my visuals and their order?	
	Am I prepared in the event that equipment will fail?	
Testing	What is the best way to determine if the audience has learned?	
	How do I plan to test or quiz them?	
Logistics	Am I familiar with the training facility?	
	How is the room arranged?	
	How will I accommodate people with special needs?	
	Can everyone see the visuals?	
Evaluations	Have I created an evaluation for the participants?	
	What is the evaluation going to measure?	
	How am I going to use the evaluation?	
	Have I done my own evaluation?	

Chapter 4
Eliciting Audience Participation

We have now covered adult learning and the different aspects of creating an effective presentation. Now it's time to discuss effective ways to present the training, specifically, ways to involve the audience. This chapter is all about different ways to interact with the participants. Remember that increased participation means a more effective training class.

This chapter will be divided into the following sections:

Conducting Introductions

This section will cover the importance of introductions and different techniques that you can use. Although introductions can be a small portion of your training, they are vital because they set the tone of the training.

Asking and Answering Questions

This section covers the basics for answering questions, basics for asking questions of the participants, and ways to use people's questions effectively. Asking and answering questions is a great way to involve people.

Dealing with Problems That Arise (Interventions)

This section talks about understanding interventions, things that can go wrong in a training class, and effectively dealing with them. Even when you spend a lot of time preparing for the training, you should always be ready for different interventions. Over the years I have rarely had a training class where an intervention did not occur, so I learned to always be prepared.

Conducting Introductions

Introductions are a key part of the effective training process. They allow the participants to feel comfortable with you, the trainer, and the other participants in the room. Once participants feel comfortable, the training experience will be more beneficial. Introductions also give you the opportunity to establish your credentials.

Introductions can also help the participants understand what they may have in common with each other, and may help you identify the key issues that you may want to address during the training.

Whether the participants know each other or not, introductions can be a great way to involve people at an early point in the training. Following are a few techniques that you can use.

Technique	How You Do It	Objective	Time (minutes)
Individual Introductions	Participants introduce themselves and explain "why I am here"	Understand the participants' backgrounds and goals	5–10
Partner Introductions	Participants introduce a "buddy"	Invite active participants	15–20
Embarrassing Moment	Participants tell an embarrassing story about themselves	Add a personal touch and say that it is okay to take risks	20–25
Favorite Activity	Participants explain their favorite non-work activity	Add interest and gain participant familiarity	20–25
True / False	Participants write 2 true statements and 1 false statement about themselves; the group votes on which is the false statement	Help learn new and different things about people that you may work with all the time; add humor	25–30
Middle Name Introductions	Participants introduce themselves using their middle name and explain how they received the name	Add humor and interest	20–25

Adapted from: Steve Ling, *Train the Trainer, PBT Fundamentals*; Catapult, Inc.; 1995 (10)

The Basics for Answering Questions

Answer Questions During the Training

I encourage participants to ask questions as they think of them for two reasons. The first reason is that they may forget the question by the end of the training. The second is that other people in the audience may have the same question. If there are too many questions (you can have a really enthusiastic group from time to time), then you can ask individuals to approach you at the end of the training.

Occasionally, you will get a question that seems out of sync with what you are discussing. Simply ask the person to bring it up again at the end of the presentation so that you and the audience are not distracted. I have a hard time answering questions on topics that were covered 5–30 minutes ago. It throws the whole class out of sync. In the same respect, I do not like to answer questions on topics that I have not covered yet.

Expect Questions

You know your presentation better than anyone else in the room, so you should think about possible questions that may arise. One key to doing this is knowing a lot about your audience. Write the expected questions and responses prior to the training so that you are prepared to answer them. Some people even have extra material they save in anticipation of additional questions.

Do Not Allow People to Dominate

Undoubtedly you will meet those individuals who actually enjoy asking multiple questions. You are not sure if they are really interested, or if they just like hearing themselves talk. Try to encourage other people to participate and always be polite. I usually say something like, "Thank you so much for your questions. Let's hear from someone else in the room now."

Do Not Let the Questioner Ramble On

This is particularly hard to control. If you feel that someone is starting to give a speech, politely interrupt them and say, "So, what you're really asking is...". If this does not work, then ask them to please state their question.

I have found that when you allow people to talk too much, they usually will stray from the topic. I taught a chemical safety class for custodial workers once, and the discussion strayed from chemical waste disposal to general waste disposal to smoking areas outside the doorways and ashes in the stairwells. When this happens, you need to politely remind them what the purpose of the training is to bring them back on track.

Listen to the Question

Sometimes it is really hard to understand what a person is trying to ask. The better you are at listening to their question, the better your response will be. Give them a chance. If they are having a hard time formulating their question, be patient and help them choose their words. Sometimes people are not sure how to say what they feel.

Repeat the Question

Repeating the question makes certain that you know exactly what they are asking. What they said, what they mean, and what you heard could all be very different things. Repeating the question also ensures that everyone in the audience has heard it. This may be difficult at first, but it makes your training clear.

Never Guess

Many times you will be asked questions for which you do not know the answers. If you do not know the answer to a question, honestly admit it. Tell them that you do not know the information right now, but that you would be happy to find out the correct response and get back to them within the next couple days. Then take down their name and phone number so that you can follow up with the correct response.

If you make up an answer, there is a good possibility you may be wrong. This means you are not only giving incorrect information, but you may also lose your credibility with the audience.

You can also try asking if anyone else in the audience knows the answer. You may have someone who is an expert and can help you. Most people are happy to help out if they know the answer.

The Basics for Asking Questions

What about asking the participants questions?

Ask Questions Simply and Sincerely

Asking questions is a great way to have the audience get involved in the training. It is also a good way to make sure that the information is sinking in. To encourage as much participation as possible, you want to make the questions simple. Always avoid sarcasm, being condescending, or making the audience feel uncomfortable. Make your questions short so that people fully understand what you are asking.

Be Patient and Wait for an Answer

This can be really tough to do, because hearing silence may be uncomfortable for you and for the audience. I have a hard time with this. If I ask a question and I do not hear a response right away, I am always tempted to answer it myself just to eliminate the silence. You have to remember that you may catch people off-guard when you ask a question, so you need to give them some time to think of an answer.

One thing that I have tried is counting to five slowly. If I do not get an answer by five, I restate the question in a different way and give them another chance to respond. Sometimes people just need to hear the question stated differently. It also gives you a chance to wake up the people who are taking a mental nap during the training.

Call on People in a Nonverbal Way

This is usually less threatening than actually using someone's name. Look one person in the eye but then ask the question to the whole group. You are inviting the person to respond, but not giving them pressure in case they do not know the correct response or they have not been paying attention.

Compliment the Person Asking the Question

By doing this, you will encourage others to ask questions and participate. Use phrases like, "You bring up a good point," "That's a great question," or "Thank you for mentioning that." It is also good to thank them for asking the question. Criticizing the person will definitely discourage people from participating. Phrases like, "Clearly you weren't listening because I just covered that" or "I'm sorry that you don't seem to understand what I'm explaining," can be very damaging in the overall training. There is nothing to be gained when you criticize someone during a training session and a great deal to be lost.

Exercise •

Exercise for Handling Questions

Here are some different situations that may happen when people are asking a question. Write down some of your ideas about how you would act. Possible answers are in Appendix I.

How Would You Respond to These Different Situations?

1. The person asking the question repeatedly interrupts you while you are trying to give an answer.

2. Someone asks a question about something that you have already covered after you have moved to a different section.

3. Someone asks a question that was already asked.

4. Someone asks a completely irrelevant question.

5. Someone asks a completely disorganized question.

6. Someone asks a question to promote him or herself.

7. Someone asks a question that only pertains to themselves.

Understanding Interventions

An intervention is an interruption of an ongoing training activity that influences the direction, content, behavior, and/or effect of the training program. It simply means that you have to do something to ensure that the training is proceeding in the way that you want. It is a way of maintaining control of the class when you feel that control is being lost.

What's Really Happening?

Understanding interventions is really important. What you think you are observing may be very different than what is actually happening. For example, let's assume that one of the participants is staring out the window. There are five different things that you, as a trainer, can assume. You can assume:

- The participant is bored because you are being too wordy and unclear
- The participant is excited because someone is streaking outside the training room
- The participant is tired because they went to bed at 4 a.m. last night
- The participant is bored because the material is too technical and complex
- The participant is distracted because they recently learned that they may lose their job

The question is, how do you know if you should intervene or not?

Effectively Dealing with Interventions

Before making an intervention, ask yourself these questions:

- Have I made any inaccurate assumptions about what indicator occurred?

  Make sure that you know exactly what facilitated the person's behavior. For example, I went to a training class once where the trainer got all upset that the audience was laughing. He thought that they were laughing at him, when it turned out that they were laughing at their goofy coworker who was making funny faces through the window of the training room.

- If I did intervene, what would I hope to accomplish?

  Always think about your purpose for intervening. For example, if you are trying to gain control of the class to ensure that everyone in the audience gains from the training as much as possible, then your motivation to intervene is good. If your motivation is to embarrass someone or make them look bad, then it is not a good motivator for intervening.

Is it low or high risk?

Low risk means that you may alienate the person who is causing trouble, but have the support of the other audience members who are being distracted by the inappropriate behavior. High risk means that you may alienate the whole audience.

I had a situation once where the most popular guy in the department, the class clown, was totally out of line. This individual poked fun at management, made loud noises with his pen, spilled a drink, and was generally rude. At one point he raised his hand and asked if he could go to the bathroom like a child. When it started to get out of hand I intervened by saying that he was welcome to leave at any time. He did not leave, he got his act together, but I alienated all his buddies who looked up to him. This was a high risk intervention but one that I felt I had to make.

Does it have a chance of succeeding?

If you do not think that your intervention is going to succeed, then you might think twice about doing it. You may ask, how do I measure success? Success means that the intervention produced the desired outcome. It may mean that you have been able to quiet a rambunctious person or encourage a quiet person to get involved in the training. If you do not think your intervention will succeed, then you might not want to do it.

Is the intervention in harmony with the ground rules that we have established?

This is an easy one. If you have established ground rules prior to the training and someone violates them, then it makes it easy for you to intervene. You can either have the participants create the rules themselves or you can create them before the training. Some of the ground rules that I have established are:

- Leave your outside issues outside class
- Respect the opinions of others
- Ask questions
- Share your experiences
- Be honest with yourself and others
- Keep an open mind and be willing to learn and try new things
- Take risks
- Participate in group activities
- Maintain confidentiality
- Do not interrupt others when they are speaking
- Return promptly from breaks

Once the ground rules are established and agreed upon, it makes it easier to intervene when they are broken.

Test Yourself •

Dealing With Problems and Interventions

These are some possible things that could go wrong during a training class. How would you respond? Some possible answers are in Appendix J.

1. Someone is sleeping in a distracting way (snoring or breathing loudly)

2. Someone has a bad attitude and is trying to cause trouble

3. Someone is rude and very confrontational

4. The group has little or no interest in the training

5. Someone comes late to the training

6. The group won't participate

7. Someone does other work while you are conducting training

Exercise •

Effectively Dealing with Interventions—Exercise #1

Would you intervene?

Evaluate the following situation and decide whether or not you would intervene and if so, how.

Situation #1:You are teaching an hour-long safety training class that covers the chemical hygiene plan and radiation safety for hospital employees working in the radiology department. This is mandatory training that all employees must attend on an annual basis, and careful records are kept. After 40 minutes of the presentation, a gentleman, the head of the radiology department, sneaks in the back door and takes a seat. Out of the corner of your eye you see him signing the attendance sheet.

What would you do? You also might want to think about what you would do if you did not know that the late person was the head of the radiology department.

When you are done, see Appendix K for a possible solution.

Exercise •

Effectively Dealing with Interventions—Exercise #2

Would you intervene?

Evaluate the following situation and decide whether or not you would intervene, and if so, how.

Situation #2:You are teaching a safety orientation for new employees that will last the entire morning (8 a.m.–noon). Before the class, one of the attendees informs you she has had the training before when the company newly employed her six months ago. Soon after she attended the orientation, she left employment for another job and has now returned. She explains that the reason that she does not want to attend is that she just finished working the night shift a half an hour ago and this is her normal sleeping time. She is also expected to report for work tonight at 11 p.m. You check with Human Resources and they say that she has to stay. After 20 minutes of the orientation, you notice that she is quietly sleeping in the last seat of the room.

What would you do? When you are done, see Appendix K for a possible solution.

Chapter 4 Summary

In conclusion, there are many different ways to elicit audience participation. This chapter specifically covers introductions, questions, and interventions.

With introductions, you basically need to remember to do them. Introduce yourself and allow people to say a little something about themselves as well. The technique that you use will depend on the size of the audience, the amount of time you have, and how well the people know each other. When conducted effectively, introductions help you to establish your credibility and create a level of comfort for the trainees.

The second section of this chapter discussed the basics for answering and asking questions as another way to involve the audience. When answering questions, be prepared, answer promptly, don't let people dominate, listen to and repeat the question, and answer only when you know the correct response. When asking the trainees questions, be simple and sincere, call on people nonverbally, and compliment the person asking the question. Handling questions effectively can make the difference between a positive or negative experience, so use the guidelines to make the most of asking and answering questions.

The final section of eliciting audience participation explained ways to effectively handle interventions. Once you understand the situation it is possible to decide whether to intervene, and if so, how. Before making an intervention, be aware of any inaccurate assumptions you may have made, know what you hope to accomplish, and decide if the intervention is low or high risk. Also, consider whether the intervention has a chance of succeeding and whether it coincides with the established ground rules.

Remember that you can never be fully prepared for every situation that may arise during training. Experience helps, but you also learn to expect anything. Encouraging people to participate makes the training more interesting, but it also means you may lose some control.

Hopefully this chapter has given you some ideas on how to involve the audience in positive ways and how to change negative situations into positive ones.

Chapter 5
Delivering Effective Training

There are three main parts to every training session: the opening, the body, and the closing. For these three sections you will want to pick a major point that you will state and restate throughout your training. Do not have more than one main point, as it may confuse people. Pick one thing that you want them to remember as they walk out that door, and make sure you keep telling them.

Opening

In the opening you explain what you are going to tell them. Pick one main point for your training and make sure that it is clear in the beginning. Here are some points that you might want to make:

- Eye protection can save your eyesight.
- An office environment may seem safe, but there are hidden hazards everywhere.
- The only safe way to enter a confined space is with a permit.
- When working with blood and body fluids, you should always assume they are contaminated.
- The Hazard Communication Program is here to make sure that you are told the proper information about the chemicals you are working with. You have the right to know.
- Always report accidents as quickly as possible.
- Protect your back when lifting, since it is the only back you will ever have.
- Don't contaminate your lungs with harmful inhalants; wear your respiratory protection.
- Even though a laser looks like a regular light beam, treat it as though it can cut through your body.

Interesting Ways to Open Your Training

- **Give a preview or outline of speech**

  "Today we're going to talk about all the reasons that you should wear fall protection."

- **Tell about some personal experience that you had**

  "I'll never forget the time I saw someone get hurt because they weren't wearing their safety glasses."

- **Ask a rhetorical question**

  "So I ask you, why are we really here? Sure, we're here for safety training, but why?"

- **Give the audience a hypothetical situation**

  "What if OSHA were to walk into your workplace tomorrow? How would you react?"

- **Tell them something new or dramatic**

  "The reason we're having this safety training is to discuss the accident that happened last week and to discuss ways that it could have been prevented."

- **Tell them a startling fact, statement, or statistic**

  "Eight out of ten adults will injure their backs during their lifetime."

- **Compliment the audience**

  "It's great to be with a group of people who I know are safety conscious and have such a low accident/incident rate."

- **Identify with the audience**

  "I know how hard it can be to remember to file an accident report on time, when what you're really concerned about is the person who got hurt."

- **Use significant or familiar quotes**

  "Safety First"

- **Use a headline or some recent news**

  "Have you heard about the recent story on CNN News that talked about a researcher dying from acute mercury poisoning?"

What To Do in the *Body* of the Training

In the body of the training, you want to restate the point and give supportive information to help them understand your point. For example, if the point you are trying to make is that eye protection is important, you could explain exactly how wearing goggles could save your eyes from harmful damage. The body is where you back up your main point with supporting information.

1. Explain what safety skills, procedures, rules, and methods they need to learn, and go into as many details as time allows.
2. Conduct any activities that will take place (e.g., practice lifting boxes for back injury prevention or fit testing a respirator for respirator safety.)
3. Ask questions to make sure that the audience is with you. For example, "Does this make sense to you? Are you following me? Am I going too fast?"
4. Use different types of material to make the class interesting and appealing to all people, and involve the participants in many ways:
    - Ask and answer questions
    - Invite comments
    - Make eye contact with each individual instead of the entire group
    - Think of games or group activities that may reinforce the training

Closing

Tell them what you told them. Restate your main point and then refer back to any specific examples that you may have given in the body. Make the closing short and sweet so they are more likely to remember it.

Interesting Ways to Close Your Training

- Give a summary of speech

  "So, in summary, remember these steps to take when you think there is a fire. Call 911, assist anyone who may need your help, and get out of the building as quickly as possible."

- Call to Action

  "I'd like to ask all of you to work real hard to remember to wear your safety glasses all the time, and to remind any coworkers who may have forgotten."

- **Predict the Future**

  "Office safety and ergonomics are very important to our society. We may even have a new OSHA Ergonomics standard to enforce within the next few years. If we have a good ergonomics policy now, then we will be ahead of the game."

- **Challenge**

  "I challenge you all to embrace the idea that behavior-based safety can make our plant a safer place to work."

- **Stirring Statement**

  "We've already had 10 serious accidents this past quarter. Who knows—the next one could involve you."

- **Quote**

  "Remember what the president of the company always says: 'Safety First.'"

- **Rhetorical Question**

  "After hearing all the information I have presented to you this morning, don't you think it is important to take the time to think before you act in an unsafe way?"

- **Bit of advice**

  "Let me tell you something. If I were you, I'd make sure that I was wearing hearing protection when I was working with noisy equipment. It's just not worth the risk."

- **Pledge or Promise**

  "I promise that increased awareness in behavior-based safety will clearly decrease our accident rates."

Timing

Utilizing time can be tricky. You can easily irritate people if you are not considerate of time constraints. Remember that their time is just as valuable as yours. Following are some helpful hints to remember.

- **Don't feel obligated to fill the entire time slot that you have been given.**

  If you cover the information in a shorter amount of time, people may appreciate the extra time to ask questions. It is always a nice treat when the training ends before planned. As a buffer you may want to advertise the session as lasting 15–30 minutes longer than you plan. This gives people a bonus when the class goes well, and it gives the instructor extra time when problems arise.

- **Try not to take more time than you are allowed.**

  This is especially important if your audience has a tight production schedule or has other obligations, such as patients that they need to see. It is always better to take less time than you are allowed. Everybody hates it when a speaker is long-winded.

- **If you intend to talk for a long time make sure you allow time for breaks.**

  It is good to allow breaks every fifty minutes. If you talk any longer than that you may make your audience uncomfortable. People appreciate the time to walk around, get some refreshments, or go to the restroom. Having breaks can clear their heads and prepare them for the next segment of training. It also gives you a break as well.

- **If you know that you are running late, make sure you tell the audience.**

  People are generally more responsive when they are told what to expect. As it gets close to the time that the session is scheduled to end, politely tell the audience that you need another 3–5 minutes to finish. Try not to take more than 5 minutes.

  It also helps to warn the audience that you are running late when you have a very enthusiastic group. I have had sessions where the participants were very vocal, asking many questions and making many comments. I normally welcome interaction, but there have been some instances where the questions have been very detailed, causing the training to run late. At that point I usually tell the audience that they have two choices: we can continue discussing questions and end the class later than planned, or they can ask their questions privately after the session is over and the training can end on time. They almost always prefer to end the session on time and ask their questions privately at the end.

- **Make sure that a clock or watch is visible.**

  Time usually passes more quickly or more slowly than you think. It is a good idea to check your watch or the clock on the wall every once in a while to make sure you are still on schedule. This is especially true if you are teaching a new class and you are unsure how long your training will last.

Chapter 5 Summary

In summary, there are three things that you have to consider when you are delivering effective training; the opening, the body and the closing. For these three parts of the training, you want to make sure you choose a main point and then restate it throughout the training.

There are different ways that you can open your training. Try to be as interesting as possible, since that will increase the chances that people will remember what your main point is. Personal experiences and headlines are just two examples of ways to open your training.

In the body of your training, you want to restate the main point and then you want to follow that with supportive information. The body of the training is usually the most detailed part and can last from 10 minutes to 5 days, depending on the length of the training class. The body is also where you would conduct any group activities or games.

The closing is your last chance to make an impression on your audience. Restate the main point and then refer to any examples that you have made in the body. As with the introduction, you want to be interesting. Clever ways of closing your training would include predicting the future, quoting someone, or offering advice.

You also need to be conscious of timing issues when delivering your training. Always be considerate of your audience and let them know if you are running late. Allow for breaks and try not to take more time than you are allowed. Finally, never feel obligated to fill the entire time slot you have been given. People usually appreciate a little extra time at the end of the session.

Chapter 6 Characteristics of Effective Presenters

How You Look and Sound is Important

Face and Eyes

Smile as much as possible when you are training. It tells the audience that you are easy to approach and that you are enjoying yourself. Remember that smiling and laughing are contagious.

Good eye contact is also crucial. It tells the audience that you care. Have you ever been to a training session or a talk where the presenter stares at the back of the room? I cannot help but think, "What is he looking at?" Looking at people tells them you are genuinely interested that they learn the material.

Good eye contact means that you look at each person in the group long enough to gain their attention, but not long enough to make them feel uncomfortable. Make sure that you also look at the people in the front and the back since it is easy to miss them. If you have notes, try not to spend too much time looking at them. Look at people instead.

Hands and Gestures

Have you ever tried to talk while sitting on your hands? It is difficult to talk without using gestures of some sort. Make sure your gestures are meaningful and that they are not distracting to the audience. Make sure you remove anything that causes you to do annoying things with your hands, like jingle change in your pocket, click a pen, or fiddle with your clothing.

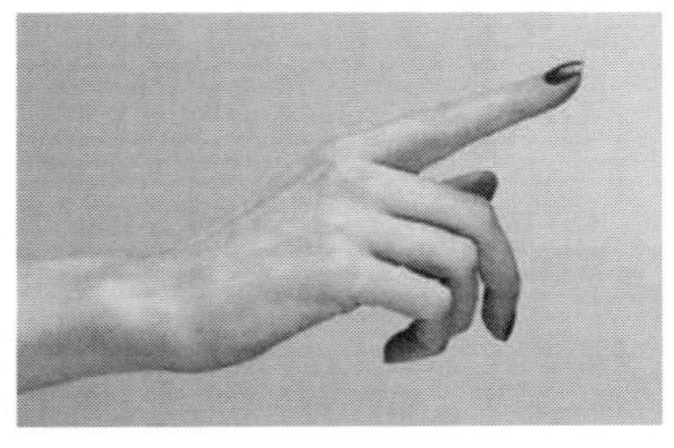

Movement and Posture

Good movement puts variation in your training. Avoid pacing back and forth, because the participants may feel like they are attending a tennis match instead of safety training. It also makes people nervous because they perceive that the trainer is nervous.

Good posture communicates confidence. Bad posture or slouching tells the audience that you are not comfortable with the material. Standing straight with your shoulders slightly back exudes confidence.

Voice

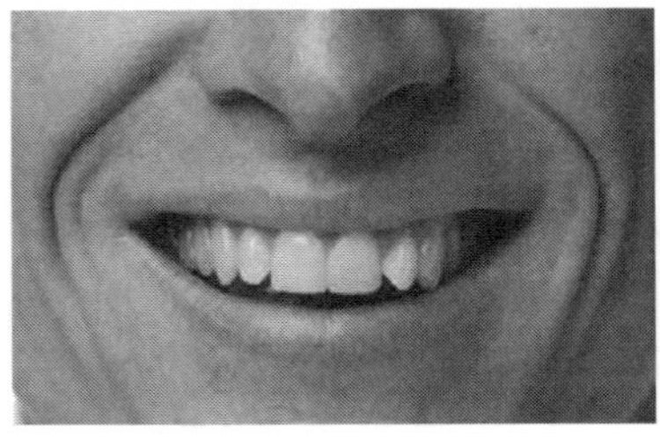

There are many things to remember about your voice. Try to avoid phrases like ah, um, and okay. This can be a hard habit to break, so you might want to ask a colleague to listen to your talk to keep track of how often you use these phrases. I'm always surprised at how often I use these phrases, even when I am consciously trying to avoid saying them.

Speak slightly faster and much louder. Try to enunciate as much as possible, especially if you are using technical words that may be difficult to say. Change your pitch often. Changing your pitch makes people pay attention, while monotone voices put people to sleep faster than reading a boring book in bed.

Remember to pause occasionally. As Artur Schnabel, a famous pianist once said, "The notes I handle no better than many pianists. But the pauses between the notes... ah, that is where the art resides!" Pauses help people to organize their thoughts and to pay attention.

Energy and Enthusiasm

Show the audience that you are excited to be there, and they will be excited to be there as well. You should also make the audience feel that you are genuinely happy to be there. No one likes a reluctant trainer. You can know your information like the back of your hand, but you may lose effectiveness without energy and enthusiasm.

Chapter 6 Summary

The final chapter of this book discusses characteristics of effective presenters. After spending so much time designing your training, it would be a shame if you could not deliver it effectively. Remember that people will be looking at and listening to you, so you should always put your best foot forward.

- Be aware of your face and eyes. Smile and laugh as much as possible and maintain good eye contact.
- Be aware of your hands and gestures. Make sure gestures have meaning and don't distract.
- Be aware of your movement and posture. Move comfortably without pacing and maintain good posture.
- Be aware of your voice. Speak loudly, enunciate, change pitch, and pause.
- Be aware of your energy level. If you are excited to be there you increase the chances that the audience will be excited, too.

Appendix A
Effective & Ineffective Training Characteristics

Effective Training Class	Ineffective Training Class
Instructor knows the material	Instructor is not well informed
Instructor is interested in the subject	Instructor doesn't seem to care
Class is held at a convenient time of the day	Class is held during a meal or at the end of the shift
Room is the proper size	Room is too large or too small
Management supports training	Management thinks training is a waste of time
Instructor listens to all questions and is able to answer them thoughtfully	Instructor answers the wrong questions or makes up incorrect answers
Instructor engages the audience	Instructor lectures without audience participation
Instructor varies voice pitch	Instructor has monotone voice
Audio-visuals are crisp and clean	Audio-visuals are messy and confusing
Room is well lit	Room is dimly lit or too bright
Room temperature is comfortable	Room temperature is too hot or too cold
Equipment works efficiently	Audio-visual equipment doesn't work
Participants understand goals and objectives	Goals and objectives are either confusing or not stated
Instructor considers the participants' training needs	Instructor teaches information that is not important to the participants and leaves out pertinent information
Everyone can see the audio-visuals	Presenter or furniture blocks participants' views
All the appropriate material is covered	Inappropriate content (too little or too much information)
Instructor allows people to comfortably ask questions	Instructor doesn't entertain questions
Instructor answers questions correctly	Instructor makes up answers
Instructor has a positive attitude while answering questions	Instructor seems inconvenienced in answering questions or is short with trainees

Effective Training Class	Ineffective Training Class
Instructions for activities are clear	Trainees are not sure what to do or how to do it
Instructor is organized and follows a clear path through the training	Instructor is confused and presents various points in an unorganized manner
One main message stated throught the training	Mixed messages
Training is stimulating for the audience	Trainins is boring

Appendix B
Answers to Adult Learning Exercise 1

Question	Answer
What is the name of the main character?	Bob
What type of vehicle does he drive?	1995 red pickup truck
What kind of saw does he operate?	Table saw
On what day of the week did the events in this story happen?	Tuesday
How long has he worked for this company?	Seven years
What time did he turn the saw on?	9:45 a.m.
What color is his spare pair of steel-toed shoes?	Black
What stone is in his high school ring?	Emerald
How old is his daughter?	Eight months
What color is his hair?	Blond

Appendix C
Answers to Adult Learning Exercise 2

Question	Answer
What color is the footrest?	Black
Where did she first want to set up her computer monitor?	In front of the window
How big is her new monitor?	Seventeen inches
What state is her office located in?	Georgia
What did Marianne set up first?	Her computer (CPU)
Where was the lever on her chair?	Bottom left side
Where did she put the CPU?	On her new carpeting on the floor
How high did she raise her seat?	Three inches
What color was her new carpeting?	Gray
Does she have a trackball or a mouse?	Trackball

Appendix D
Crossword Puzzles

Chemical Safety Crossword Puzzle

Across

5. Chemical known or suspected of causing cancer.
6. Minimum temperature at which liquid gives off enough vapor to ignite.
9. Alter an acid or base so its pH is 7.
11. Chemical with a pH greater than 7.
12. Solid or liquid that can spontaneiously ignite.

Down

1. Chemical that causes development of allergic reaction in normal tissue.
2. Ease with which a liquid, solid, or gas ignites, spontaneously or as result of spark or flame.
3. Chemical with a pH less than 7.
4. "CHP" stands for Chemical ___ Plan.
7. Chemical compound that induces DNA mutations.
8. Any substance harmful to living tissue.
10. "TWA" stands for ___ Weighted Average.

Answers to Chemical Safety Crossword Puzzle

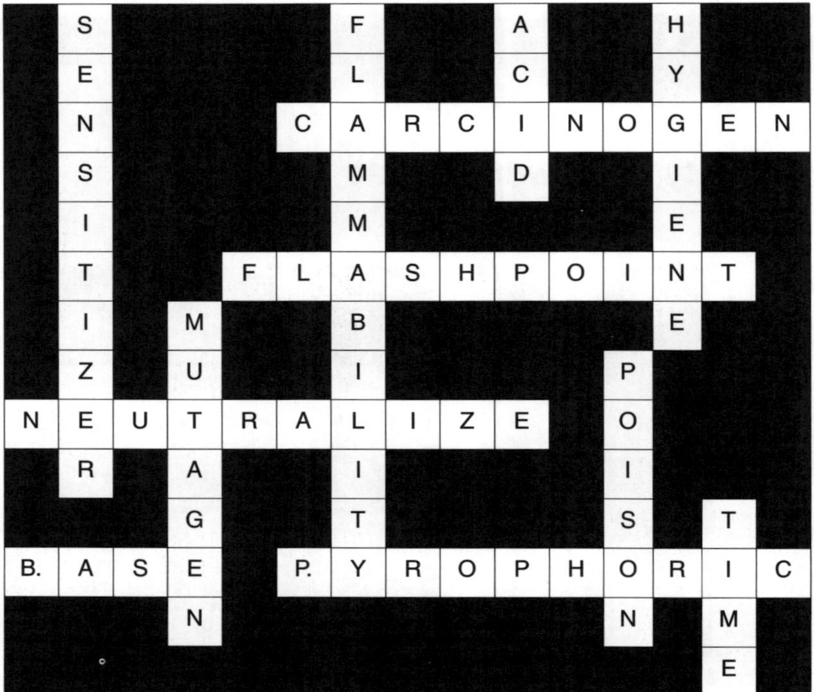

Laser Safety Crossword Puzzle

Across

4. "MPE" stands for Maximum ___ Exposure Limit.
7. Lasers have three properties: they are monochromatic, coherent, and ___.
8. Laser damage to this part of the eye can cause a loss in vision.
9. Gives energy in single or multiple bursts.
11. The most dangerous type of laser is this class number.

Down

1. Type of laser that emits a steady stream of light.
2. "LASER" is an acronym for Light ___ by Stimulated Emission of Radiation.
3. Electromagnetic radiation with wavelengths from 180–400 nm.
5. The most common non-beam hazard.
6. The most important type of personal protective equipment when working with lasers.
10. Laser damage to this part of the eye can cause a "milky area" or cataract to form.

Answers to Laser Safety Crossword Puzzle

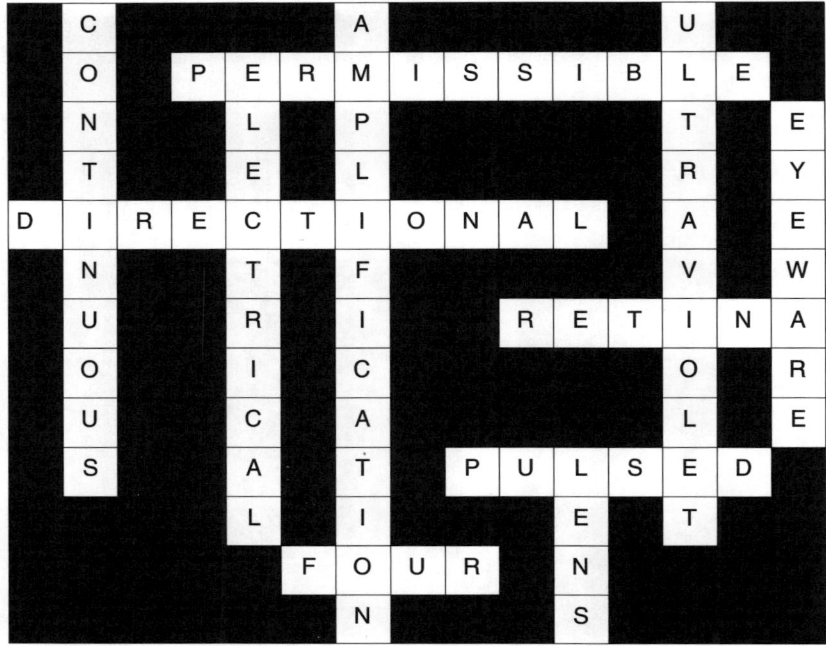

Hearing Conservation Crossword Puzzle

Across

2. "STS" stands for Standard ___ Shift.
5. Audiogram against which future audiograms are compared.
8. Audiometric test showing person's hearing ability at different frequencies.

Down

1. Unit of measurement of frequency.
3. Noise ___ is an instrument that measures a noise dose.
4. "TWA" stands for Time ___ Average.
6. Type of personal protective equipment for hearing conservation.
7. How often instruments are calibrated.
8. ___ level is an 8-hour TWA of 85 dB.
9. Unit of measurement of sound level.

Answers to Hearing Conservation Crossword Puzzle

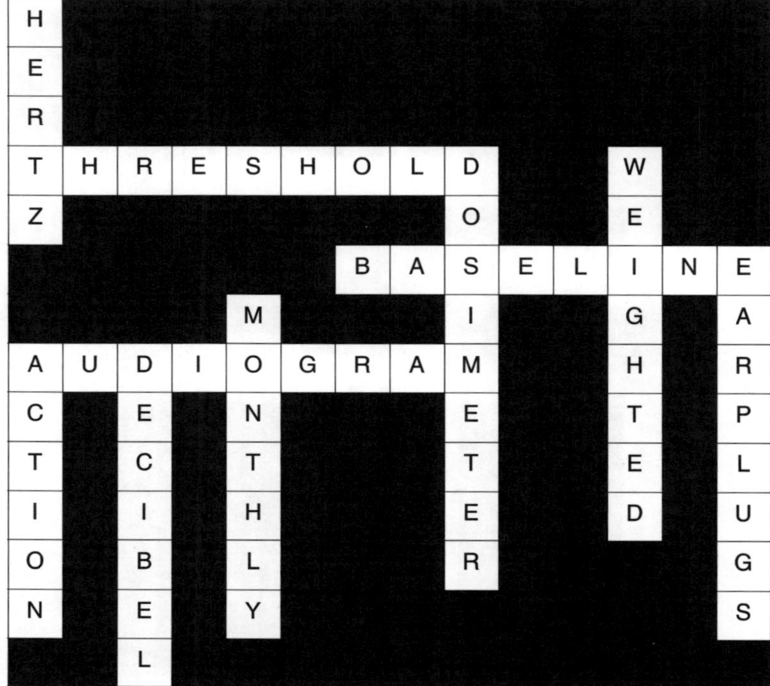

Appendix E
Ergonomics Jeopardy

Ergonomics Jeopardy Gameboard

Lighting	Cumulative Trauma Disorders (CTD)	Lifting	Computer Workstations
100	100	100	100
200	200	200	200
300	300	300	300

Ergonomics Jeopardy Questions

Topic	Points	Question	Answer
Lighting	100	One thing the quality of light is determined by.	What is the color of light, direction of light, or amount of glare?
	200	When levels fall below this amount, workers can have a negative reaction to the lighting.	What is 20 foot-candles?
	300	One of two major sources of light in industrial settings.	What is daylight or electrical?
Culumative Trauma Disorders	100	A common CTD for office assistants.	What is carpal tunnel syndrome?
	200	One key reason for increasing numbers of CTD.	What is an increase in computer use and automation, or the aging workforce?
	300	One way to decrease trigger finger.	What is designing tool handles that operate with the thumb?
Lifting	100	Part of your body you want to use when lifting.	What are your legs?
	200	What to do if your load is too heavy to lift.	What is get help from a coworker, or use a mechanical lifting device?
	300	Avoid moving your body this way when you are lifting.	What is twisting or turning?

Topic	Points	Question	Answer
Computer Workstations	100	When looking at your monitor, try to avoid this by not placing your monitor by a window.	What is glare?
	200	Chairs should be equipped with these.	What are back rests or lumbar support?
	300	Angle between the upper arm and forearm when using a key-board.	What is 70–90 degrees?

Appendix F
Creating Patterns for Training

Pattern	Type of Training	Example
Problem / Solution	Hazard Communication	During our lab inspections last month, we found 33 unlabeled chemicals. Let's discuss the Hazard Communication program and what type of information needs to be on a label.
Chronological	Lock Out / Tag Out	Three years ago we had one violation, two years ago we had four violations, and this year we had six. Since the violations are increasing each year, let's revisit the lock out / tag out policy so that next year our violations will be the lowest yet.
Physical Location	Hazard Communication	Today we are going to talk about the Hazard Communication program. I am going to divide my talk into the three floors of our plant since the chemicals are different for each floor. You need to be familiar with the different chemicals since you work on all three floors.
Extended Metaphor or Analogy	Personal Protective Equipment	Remember how you felt on your first day of school? You were excited because you knew you were going to learn things that would make you more intelligent, but you were also intimidated because you were going to have to work hard. It's the same way with personal protective equipment. Taking the time to make sure you use the equipment properly may be difficult but it will only help you in the long run.
Divide a Quote	Back Injury Prevention	"Lift with your legs." Let's first talk about some lifting techniques that you should use. Then we'll talk about the ways that you can save your body from injury. Finally, we'll demonstrate some safe and effective ways to use your legs when lifting.

Pattern	Type of Training	Example
Divide a Word	Ergonomics	**E** stands for **Energy.** You'll have more energy if you adapt your workplace to your body, rather than adapting your body to your workplace. **R** stands for **Rearrange.** Rearrange your office equipment for optimum comfort before you start working. **G** stands for **Good chairs.** Chairs can make the difference between a happy or painful back. **O** stands for **One size does not fit all.** Adjust your chair to fit your body. **N** stands for **Noise.** Noise can cause fatigue. Decreasing your noise level can keep you more alert. **O** stands for **Optimizing light.** The lighting should be slightly dimmer than usual office lighting. **M** stands for **Moving your keyboard.** Move your keyboard so that it is at elbow height. **I** stands for **Increasing your comfort while you work.** **C** stands for **Changing your tasks often.** Alternating your tasks can decrease your tension. **S** stands for **Stretching.** Stretch breaks help to reduce muscle tension—and it feels good, too!

Appendix G
Selecting a Training Method

Safety Training	Possible Training Methods
Accident Investigation	Discussion—Discuss various accidents and ways they were investigated. Case Studies—Present an accident and have the participants try to analyze how they would go about investigating it.
Back Injury Prevention	Demonstration—Demonstrate proper lifting techniques or proper ways to wear a back belt. Stretching Exercises—Have participants practice some different exercises (toe touches, chin tucks, etc.).
Lab Safety	Role Playing—Have participants act out what they would do in the event of a chemical spill. In-Basket—Drop some chemical safety questions into a hat and have groups choose a question and answer it.
Lock Out / Tag Out	Exercise—Bring some small equipment and ask the participants to practice locking and tagging the equipment out for various situations. Games—Create a Lock Out / Tag Out crossword puzzle.
New Employee Orientation	Lecture—Present the safety information to a large group of new employees. Case Study—Present different situations to the audience (e.g. seeing smoke come from a doorway) and ask them to explain the proper procedures.
Respiratory Protection	Demonstration—Fit testing is an excellent form of demonstration because participants learn what it feels like to have a properly-fitting respirator. Brainstorming—Ask the participants to share three different situations where they need to wear respiratory protection in their present job.

Safety Training	Possible Training Methods
First Aid	Demonstration—Demonstrate the proper way to bandage a wound. Case Study—Give the audience various examples of when they would have to use first aid and what they would do.
Hearing Conservation	Discussion—Discuss the importance of hearing protection. Demonstration—Demonstrate the proper way to insert ear plugs.
Materials Safety Data Sheet Training	Lecture—Explain the different sections of a Material Safety Data Sheet and the information that is contained in each section. Structured Exercise—Hand out Material Safety Data Sheets for different chemicals that the participants use and ask them to find different information (e.g. the hazards, first aid, or proper handling).
Confined Space	Discussion—Discuss what different environments constitute a confined space. Case Study—Explain the different situations where employees had to enter confined space and ask the participants to determine the proper procedures for entering the various confined spaces.
Cardiopulmonary Resuscitation (CPR)	Lecture—Explain the different steps to take in performing CPR. Demonstration—Physically show the class how to perform CPR.
Radiation Protection	Case Study—Describe a hypothetical or actual radiation exposure and ask the class to discuss ways that it could have been prevented. Role Playing—Have members represent different people (Nuclear Regulatory Commission officer, radiation protection officer, radiation waste collector, lab principal investigator, lab technician, post-doctoral fellow, and graduate student) and have them act out their roles when using radioactive materials.
Asbestos Awareness	Discussion—Discuss the various hazards of asbestos exposure. Brainstorming—Brainstorm on the different ways that asbestos was used in the past and the best method for removal.

Safety Training	Possible Training Methods
Ergonomics	Demonstration—Demonstrate ways to set up your computer workstation for optimum comfort. Case Study—Show the participants bad ergonomic situations and ask them how they would improve them.
Bloodborne Pathogens	Lecture—Explain the different bloodborne pathogens and why they are hazardous. Exercises—Have participants practice the proper way to remove soiled gloves.

Appendix H
Pros and Cons of Different Types of Media

Media Type	Pros	Cons
Videotape	• You can show things that you might not be able to otherwise demonstrate. • May increase audience's attention when used effectively.	• Material may not be entirely pertinent to your audience. • May be costly.
Handout	• Great way to give the participants something to take back with them. • Helps reinforce taught messages.	• Participants may lose handouts or may not read them. • Time consuming to put together and expensive, depending on the size of the audience.
Overhead Projector / Transparencies	• You can write on them. • Easily viewed in lighted room. • Easy to use projector and to make overheads. • Great for different-sized groups.	• Print may be too small. • Bulb could burn out. • Machine is sometimes noisy.
Laptop Computer and LCD Projector	• Interactive if you use presentation software. • Easily changed for each presentation, unlike slides or overheads.	• Very expensive. • Not very portable. • Potential for technical problems.
Flip Chart	• Portable. • Great for brainstorming. • Versatile—you can prepare them before the presentation or use them to write ideas during discussion.	• Could be intimidating for the person doing the writing. • Not a permanent record of information unless you transfer information to a handout.
Slides and Slide Projector	• Effective for showing photographs. • Equipment is small and easy to move around.	• Can be expensive to make. • Slides are permanent—you can't make any changes once they are created.
Actual Object	• Shows people what you are talking about 'hands-on."	• Take the risk that it may be damaged.

Media Type	Pros	Cons
Computer Based Training	• One-on-one safety training: employees take training at their own pace. • Convenient—employees can take training whenever they want. • Professional and impressive.	• Very expensive. • Bad option for people not comfortable with computers. • Customizing takes time and money. • No human interaction. • May need to instruct participants how to use computer and software.

Appendix I
Exercise for Handling Questions

How would you respond to these different situations?

1. **The person asking the question repeatedly interrupts you while you are trying to give an answer.**

   I have a difficult time when people interrupt me. I usually let them do it once and then I politely smile and say, "Excuse me, please let me finish." I keep saying that every time they interrupt. Eventually they will stop.

2. **Someone asks a question about something that you have already covered after you have moved to a different section.**

   It depends on how vital the question is. If it is something that I think will benefit the group or something that I forgot to mention, then it is worth answering at this point. If it is a detailed question that is not important and will take a long time for me to answer, I usually say something like, "Oh, so you are referring to the last section. Let me talk to you more about it after the class."

3. **Someone asks a question that was already asked.**

   When it happens the first time I usually just answer it again. If it happens a second time I usually say, "As I said earlier…". Then I might follow it with, "Does that make sense?"

4. **Someone asks a completely irrelevant question.**

   This usually happens when there is a hidden agenda in the group. I used to have management and personnel questions come up during safety training because that is the only time that some people feel they can vent their feelings. I usually respond by saying, "I see how that would concern you, but that is really not what we are talking about today. That might be a good question for your manager."

5. **Someone asks a completely disorganized question.**

   I try to bring them back to center by restating, in my words, what I think they are trying to say. I might say, "I think I understand where you are coming from. So, what you're really asking is…." Sometimes I am not able to restate their question. If this happens then I say, "I'm sorry, I'm not sure what your question is." or "what exactly are you asking me?" You'll find that sometimes they don't have questions at all. They're just trying to vent some feelings.

6. **Someone asks a question to promote him- or herself.**

   This is another tough one. I never mind if someone says something to promote themselves, but if it rambles on or happens more than once, then I usually cut them off and say something like, "That's wonderful, but let's concentrate on the group today."

7. **Someone asks a question that only pertains to themselves.**

   If I am certain that it only concerns that one person, then I ask him or her to approach me at the end of the training. I might say something like "That's a good question but that's really a special situation in your case. Come see me after the training and I'll be happy to answer your question then." You may lose the rest of the audience if you discuss situations that don't concern them, so try to address this type of question individually. You can approach the person during a break or at the end of training.

Appendix J
Dealing With Problems and Interventions

These are some things that can go wrong during a training class. How would you respond?

1. **Someone is sleeping in a distracting way (snoring or breathing loudly)**

   I find sleepers very distracting, especially when they are loud. Sometimes I think they are more distracting to me than to the other people in the audience. The first thing I do is slowly walk towards the person. As I get closer, my voice gets louder and that is usually enough to wake them up. If that doesn't work, a friendly tap on the shoulder or their foot is enough to gently wake them.

2. **Someone has a bad attitude and is trying to cause trouble**

   Depending on how badly they are behaving, you may have to ask them to leave. I have only threatened to do that a few times. Usually the threat is enough to make people realize that they need to calm down and act more professionally. No one needs a disruptive troublemaker.

3. **Someone is rude and very confrontational**

   Again, you can ask them to leave. You can also be honest and tell them that you don't appreciate their tone of voice or the way that they are acting. I once had someone question my credentials in the beginning of a training class. He claimed that I didn't look more than sixteen and that I shouldn't be teaching him about chemical safety. I very slowly told him all the details of my undergraduate and graduate work in chemistry and then told him about my experience training over 10,000 employees in chemical safety. He was quiet and cooperative after that.

4. **The group has little or no interest in the training**

   That's a tough one. You definitely have to motivate them early on. I find that the biggest motivation in safety training is that safe work will maintain health and happiness in their personal lives. That's not to say that they won't be happy without the training, but they sure won't be happy when they have that radiation exposure or spend some time out of work because they have thrown their back out. I also try to remind people that what you do at work carries over to your personal life. You may break your leg at work, but that means that you won't have much fun playing with your children or grandchildren or trying to do the things that you enjoy in your leisure time.

Always try to establish motivation early and that usually will help them maintain interest. Also try a variety of material so they always have something different to listen to.

5. Someone comes late to the training

It depends on how late they are. If they show up five to ten minutes late then that's fine. If they miss more than half of the training, I usually ask them to come back to another training session and I make sure that they have the schedule. Letting them stay isn't fair to the people who have come for the entire session and it may not fulfill the OSHA training requirements.

6. The group won't participate

Again, motivation is key. I like to use candy a lot. It sounds childish but it works great. It gives them positive rewards for their efforts. The first time that I did a crossword puzzle was rough. People were trying to be polite but I could tell that they weren't into it. The next time I brought little snack-size candy bars and I couldn't believe the difference. People raced through the puzzle and had a great time teasing the other teams. Whenever I use candy or any other reward I make sure that I reward everyone in some way.

7. Someone does other work while you are conducting training

I've had people read newspapers, grade papers, and work on reports during my training. Believe it or not, I actually had a person come to training with a Walkman on. When this happens I politely ask them to put the materials away until after the training is done. If they still don't put it away then I approach them a second time and tell them that I don't think it's a good idea that they sign their name to the attendance list since they are not paying attention to the training. This usually encourages them to put their materials away.

Appendix K
Would You Intervene?

Effectively Dealing with Interventions

Exercise #1

Situation: You are teaching an hour-long safety training class that covers the chemical hygiene plan and radiation safety for hospital employees working in the radiology epartment. This is mandatory training that all employees must attend on an annual basis and careful records are kept. After 40 minutes of the presentation, a gentleman, the head of the radiology department, sneaks in the back door and takes a seat. Out of the corner of your eye you see him signing the attendance sheet.

In this situation, I politely asked him to leave. Luckily, I did not know that he was the head of the radiology department at the time, but I certainly was aware of that fact when I returned to my office.

There were a couple of reasons that I asked him to leave. First of all, he had missed the majority of the training and I couldn't say that he had completed the training in good faith. Secondly, it set a horrible example for the rest of the people in the audience. If I had let him stay then they might have thought it was acceptable for them to come late also. Finally, this was one of four or five different classes so I knew that he could easily take the training at another time.

Exercise #2

SItuation: You are teaching a safety orientation for new employees that will last the entire morning (8 a.m. - noon). Before the class, one of the attendees informs you she has had the training before when the company newly employed her six months ago. Soon after she attended the orientation, she left employment for another job and has now returned. She explains that the reason that she does not want to attend is that she just finished working the night shift a half an hour ago and this is her normal sleeping time. She is also expected to report for work tonight at 11 p.m. You check with Human Resources and they say she has to stay. After 20 minutes of the orientation, you notice that she is quietly sleeping in the last seat of the room.

In this situation, I let her sleep for a couple of reasons. First of all, I remembered her from the training six months ago. I knew that my training had not changed since then, so I was fairly certain that she had already received the proper training. Secondly, she was not being disruptive in the back of the room and I saw no reason to wake her up. Management was not respecting her time by asking her to attend training when she was off work, so I felt that I should give her some extra consideration.

Additional Reading

Brookfield, Stephen. *Understanding and Facilitating Adult Learning*. San Francisco: Jossey-Bass Publishers, 1986.

Bullard, Rebecca; Brewer, Mary Jean; Gaubas, Nancy; Gibson, Angie; Hyland, Kathy and Sample, Eileen. *The Occasional Trainer's Handbook*. Englewood Cliffs, NJ: Educational Technology Publications, 1994.

Charney, Cyril. *The Trainer's Tool Kit*. New York: Amacom, 1998.

Craig, Robert L. *The ASTD Training and Development Handbook*. McGraw-Hill, 1996.

Dirkx, John M. and Prenger, Suzanne M. *A Guide for Planning and Implementing Instruction for Adults: A Theme-Based Approach*. San Francisco: Jossey-Bass Publishers, 1997.

Hayes, Elisabeth. *Effective Teaching Styles*. San Francisco: Jossey-Bass Publishers, 1989.

James, Roger. *The Techniques of Instruction*. Brookfield, VT: Gower, 1995.

Knox, Alan B. *Teaching Adults Effectively*. San Francisco: Jossey-Bass Publishers, 1980.

Mayo, Douglas G. and DuBois, Phillip H. *The Complete Book of Training: Theory Principles, and Techniques*. San Diego: University Associates, Inc., 1987.

Monkhouse, Bob. *Just Say a Few Words: The Complete Speaker's Handbook*. New York: M. Evans and Co., 1991.

Nilson, Carolyn. *Training and Development Yearbook 1996/1997*. Englewood Cliffs, NJ: Prentice Hall, 1996.

Pike, Bob. *Creative Training Tools*. Minneapolis, MN: Lakewood Books, 1994.

———. *Managing the Front End of Training*. Minneapolis, MN: Lakewood Books, 1994.

———. *Motivating Your Trainees*. Minneapolis, MN: Lakewood Books, 1994.

———. *Optimizing Training Transfer*. Minneapolis, MN: Lakewood Books, 1994.

———. *Powerful Audiovisual Techniques*. Minneapolis, MN: Lakewood Books, 1994.

Powers, Bob. *Instructional Excellence: Mastering the Delivery of Training*. San Francisco: Jossey-Bass Publishers, 1992.

Rothwell, William J. and Kazanas, H.C. *Mastering the Instructional Design Process.* San Francisco: Jossey-Bass Publishers, 1992.

Silberman, Melvin L. *101 Ways to Make Training Active.* San Diego, CA: Pfeiffer, 1995.

———. *Active Training: A Handbook of Techniques, Designs, Case Examples, and Tips.* Lexington, Massachusetts: Lexington Books, 1990.

Simmons, Sylvia H. *How to be the Life of the Podium: Openers, Closers, and Everything in Between to Keep Them Listening.* New York: Amacom, 1991.

Smith, Robert M. *Helping Adults Learn How to Learn.* San Francisco: Jossey-Bass Publishers, 1983.

Smith, Terry C. *Making Successful Presentations: A Self-Teaching Guide.* New York: Wiley, 1991.

Sork, Thomas J. *Designing and Implementing Effective Workshops.* San Francisco: Jossey-Bass Publishers, 1984.

Sprague, Jo. *The Speaker's Handbook.* Fort Worth: Harcourt Brace College Publishers, 1996.

Stanley, Lloyd A. *Guide to Evaluation of Training.* Ljubljana, Yugoslavia: International Center for Public Enterprises in Developing Countries, 1987.

———. *Guide to Training Need Assessment.* Ljubljana, Yugoslavia: International Center for Public Enterprises in Developing Countries, 1987.

———. *Training Curriculum Development.* Ljubljana, Yugoslavia: International Center for Public Enterprises in Developing Countries, 1987.

Sullivan, Richard L. *Technical Presentation Workbook: Winning Strategies for Effective Public Speaking.* New York: ASME Press, 1996.

Tomlinson, Gerald. *Speaker's Treasury of Sports Anecdotes, Stories and Humor.* Englewood Cliffs, NJ: Prentice Hall, 1990.

Wilson, John P. *Materials for Teaching Adults: Selection, Development and Use.* San Francisco: Jossey-Bass Publishers, 1983.

Inspirational Quotes on Public Speaking

Talkers have always ruled. They will continue to rule. The smart thing is to join them.

—Bruce Barton

Faced with the choice between changing one's mind and proving there is no need to do so, almost everyone gets busy on the proof.

—John Kenneth Galbraith

Give us the tools and we will finish the job.

—Winston Churchill

What you do speaks so loud I can't hear what you say.

—Ralph Waldo Emerson

An eye can threaten like a loaded and leveled gun; or can insult like hissing and kicking; or in its altered mood by beams of kindness, make the heart dance with joy.

—Ralph Waldo Emerson

Stand tall. The difference between towering and cowering is totally a matter of inner posture. It's got nothing to do it with height, it costs nothing and it's more fun.

—Malcolm Forbes

You never get a second chance to make a good first impression.

—John Molloy

Perhaps of all the creations of man, language is the most astonishing.

—Lytton Strachey

The notes I handle no better than many pianists. But the pauses between the notes... ah, that is where the art resides!

—Artur Schnabel

Your listeners won't care how much you know until they know how much you care!

—Anonymous

The one who causes them to laugh, gains more votes for the measure than the one who forces them to think.

—Malcolm de Chazall

Laughter is the shortest distance between two people.

—Victor Borge

Grow antennae, not horns.

—James Burrill Angell, President of the University of Michigan

Nothing great was ever achieved without enthusiasm.

—Ralph Waldo Emerson

The ability to express an idea is well nigh as important as the idea itself.

—Bernard Baruch

Nothing is so contagious as enthusiasm. It moves stones. It charms brutes. Enthusiasm is the genius of sincerity, and truth accomplishes no victories without it.

—Edward George Earle Bulwer-Lytton

The man who can think and does not know how to express what he thinks is at the level of him who cannot think.

—Pericles

Whether you think you can or you think you can't, you're probably right.

—Anonymous

Do the act and the attitude follows.

—William James

The skill to do comes from the doing.

—Cicero

He who risks nothing need hope for nothing.

—Friedrich Schiller

Each time we ask more of ourselves than we think we are able to give, and then manage to give it, we grow.

—Anonymous

Index

V

W